Studies in Computational Intelligence

Volume 926

Series Editor

Janusz Kacprzyk, Polish Academy of Sciences, Warsaw, Poland

The series "Studies in Computational Intelligence" (SCI) publishes new developments and advances in the various areas of computational intelligence—quickly and with a high quality. The intent is to cover the theory, applications, and design methods of computational intelligence, as embedded in the fields of engineering, computer science, physics and life sciences, as well as the methodologies behind them. The series contains monographs, lecture notes and edited volumes in computational intelligence spanning the areas of neural networks, connectionist systems, genetic algorithms, evolutionary computation, artificial intelligence, cellular automata, self-organizing systems, soft computing, fuzzy systems, and hybrid intelligent systems. Of particular value to both the contributors and the readership are the short publication timeframe and the world-wide distribution, which enable both wide and rapid dissemination of research output.

The books of this series are submitted to indexing to Web of Science, EI-Compendex, DBLP, SCOPUS, Google Scholar and Springerlink.

More information about this series at http://www.springer.com/series/7092

Luefeng Chen · Min Wu · Witold Pedrycz ·
Kaoru Hirota

Emotion Recognition and Understanding for Emotional Human-Robot Interaction Systems

Springer

Luefeng Chen
School of Automation
China University of Geosciences
Wuhan, China

Min Wu
School of Automation
China University of Geosciences
Wuhan, China

Witold Pedrycz
Department of Electrical
and Computer Engineering
University of Alberta
Edmonton, AB, Canada

Kaoru Hirota
Tokyo Institute of Technology
Yokohama, Japan

ISSN 1860-949X ISSN 1860-9503 (electronic)
Studies in Computational Intelligence
ISBN 978-3-030-61576-5 ISBN 978-3-030-61577-2 (eBook)
https://doi.org/10.1007/978-3-030-61577-2

This Springer imprint is published by the registered company Springer Nature Switzerland AG
The registered company address is: Gewerbestrasse 11, 6330 Cham, Switzerland

Preface

As robots enter into every aspect of daily life, people put forward higher requirements on robots, hoping that robots come with abilities to perceive human emotions and intentions. Such robots are called emotional robots. Their emergence will change the traditional human-robot interaction mode and realize the emotional interaction between humans and robots. Emotional robot is to use artificial intelligence methods and technologies to endue robots with human-like emotions, so that they have the ability to recognize, understand, and express joy, sorrow, and anger. The robot revolution has entered the era of "Internet + emotion + intelligence". In the facc of the urgent demand for emotional robots in the domestic and foreign markets, it is indispensable to break through the key technologies of human-robot interaction and emotional computing. Therefore, promoting the intelligent robots, make them sense the surrounding environment, to understand human emotion, intention and service demand, adaptively realize human-robot interaction with users, according to the needs of users and the change of environmental information to provide high quality service, has become the trend of the developments of a new generation of intelligent robots. Such development exhibit important research significance and evident application value. Aiming at the development needs of emotional robots and human-robot emotional interaction systems, this book introduces the fundamental concepts, system architecture, and system functions of emotional computing and emotional robot systems. The book focuses on the key technologies and scientific problems involved in the emotional robot system, such as multimode emotion recognition and emotion intention understanding, and presents the design and application examples of human-robot emotional interaction system.

This book is organized into 13 chapters. Chapter 1 introduces the basic knowledge of multimodal emotion recognition, emotional intent understanding, and emotional human-robot interaction system, and explains the complete process of emotional human-robot interaction. In Chap. 2, combined with the characteristics of facial expression, speech, and gesture, the construction method of multimodal emotional feature set is systematically described. In Chap. 3, Softmax regression-based deep sparse autoencoder network is proposed to recognize facial emotion. Chapter 4 introduces AdaBoost-KNN using adaptive feature selection with direct optimization is proposed for dynamic emotion recognition. In Chap. 5, the weight-adapted convolution neural network is proposed to extract discriminative expression representations for recognizing facial expression. Chapter 6 presents the two-layer fuzzy multiple random forest for speech emotion recognition. In Chap. 7, the two-stage fuzzy fusion based-convolution neural network is presented for dynamic emotion recognition by using both facial expression and speech modalities. Chapter 8 presents the Dempster-Shafer theory based on multi-SVM to deal with multimodal gesture images for intention understanding. In Chap. 9, the three-layer weighted fuzzy support vector regression model is proposed for understanding human intention, which is based on the emotion-identification information in human-robot interaction. Chapter 10 proposes a two-layer fuzzy support vector regression-Takagi-Sugeno model for emotion understanding. In Chap. 11, an intention understanding model based on two-layer fuzzy support vector regression is proposed in human-robot interaction. Chapter 12 introduces the basic construction method of emotional human-robot interaction systems based on multimodal emotion recognition and emotion intention understanding. In Chap. 13, simulation experiments and application results of our emotional human-robot interaction system are shown.

We are grateful for the support of the National Natural Science Foundation of China under Grants 61973286, 61603356, 61210011, and 61733016, the 111 project under Grant B17040, and the Fundamental Research Funds for the Central Universities, China University of Geosciences (No. 201839).

We are also grateful for the support of scholars both at home and abroad. We would like to thank Prof. Jinhua She of Tokyo University of Engineering, Profs. Yong He, Weihua Cao, and Xin Chen, Assoc. Prof. Zhentao Liu of China University of Geosciences for their valuable help. Finally, we would like to express our appreciation for the great effort of graduate students Min Li, Kuanlin Wang, Wuanjuan Su, Yu Feng, Wenhao Duan, Pingping Zhang, and Wei Cao of China University of Geosciences.

Wuhan, China
August 2020

Luefeng Chen
Min Wu
Witold Pedrycz
Kaoru Hirota

Contents

1 Introduction 1
1.1 Emotion Feature Extraction and Recognition 3
1.2 Emotion Understanding 8
1.3 Emotional Human-Robot Interaction System 9
1.4 Organization of This Book 10
References 11

2 Multi-modal Emotion Feature Extraction 15
2.1 Introduction 15
2.2 Facial Expression Feature Extraction 17
2.3 Speech Emotion Feature Extraction 18
2.4 Gesture Feature Extraction 19
2.5 Summary 22
References 22

3 Deep Sparse Autoencoder Network for Facial Emotion Recognition 25
3.1 Introduction 25
3.2 Softmax Regression Based Deep Sparse Autoencoder Network 27
3.3 ROI Based Face Image Preprocessing 29
3.4 Expand the Encode and Decode Network 30
3.5 Softmax Regression 31
3.6 Overall Weight Training 32
3.7 Experiments 33
3.7.1 Fine-Tune Effect on Performance of Recognition 34
3.7.2 The Number of Hidden Layer Node's Effect on Performance of Recognition 36
3.7.3 Recognition Rate 37

3.8 Summary 37
References 38

4 AdaBoost-KNN with Direct Optimization for Dynamic Emotion Recognition 41
4.1 Introduction 41
4.2 Dynamic Feature Extraction Using Candide-3 Model 43
4.3 Adaptive Feature Selection Based on Plus-L Minus-R Selection 44
4.4 AdaBoost-KNN Based Emotion Recognition 45
4.5 AdaBoost-KNN with Direct Optimization for Emotion Recognition 47
4.6 Experiments 48
4.6.1 Experimental Environment and Data Selection 48
4.6.2 Simulations and Analysis 49
4.6.3 Preliminary Application Experiments on Emotional Social Robot System 52
4.7 Summary 54
References 54

5 Weight-Adapted Convolution Neural Network for Facial Expression Recognition 57
5.1 Introduction 57
5.2 Facial Expression Image Preprocessing 59
5.3 Principal Component Analysis for Extracting Expression Feature 60
5.4 Weight-Adapted Convolution Neural Network for Recognizing Expression Feature 60
5.4.1 Feature Learning Based on Deep Convolution Neural Network 61
5.4.2 Softmax Regression for Feature Recognition 62
5.5 Hybrid Genetic Algorithm for Optimizing Weight Adaptively 62
5.6 Experiments 64
5.7 Summary 71
References 72

6 Two-Layer Fuzzy Multiple Random Forest for Speech Emotion Recognition 77
6.1 Introduction 77
6.2 Feature Extraction 79
6.3 Fuzzy-c-Means Based Features Classification 81
6.4 Two-Layer Fuzzy Multiple Random Forest 82

6.5 Experiments 82
6.5.1 Data Setting 83
6.5.2 Environment Setting 83
6.5.3 Simulations and Analysis 84
6.6 Summary 87
References 88

7 Two-Stage Fuzzy Fusion Based-Convolution Neural Network for Dynamic Emotion Recognition 91
7.1 Introduction 91
7.2 Dynamic Emotion Feature Extraction 93
7.3 Deep Convolution Neural Network for Extracting High-Level Emotion Semantic Features 95
7.4 Feature Fusion Based on Canonical Correlation Analysis 96
7.5 Decision Fusion Based on Fuzzy Broad Learning System 97
7.6 Two-Stage Fuzzy Fusion Strategy 101
7.7 Experiments 101
7.7.1 Data Setting 101
7.7.2 Experiments for Hyperparameters 102
7.7.3 Experimental Results and Analysis 107
7.8 Summary 112
References 113

8 Multi-support Vector Machine Based Dempster-Shafer Theory for Gesture Intention Understanding 115
8.1 Introduction 115
8.2 Foreground Segmentation and Feature Extraction 117
8.2.1 Foreground Segmentation Based on Depth and RGB Images 117
8.2.2 Speeded-Up Robust Features Based Gesture Feature Extraction 118
8.3 Encoding Speeded-Up Robust Features: Sparse Coding 119
8.4 Multi-class Linear Support Vector Machines 119
8.5 Dempster-Shafer Evidence Theory for Decision-Level Fusion 120
8.6 Experiments 122
8.6.1 Experimental Setting 122
8.6.2 Experimental Environment and Setup 124
8.6.3 Experimental Results and Analysis 124
8.7 Summary 129
References 129

9 Three-Layer Weighted Fuzzy Support Vector Regressions for Emotional Intention Understanding 133
9.1 Introduction 133
9.2 Support Vector Regression 136
9.3 Three-Layer Fuzzy Support Vector Regression 137
9.4 Characteristics Analysis of Emotional Intention Understanding 140
9.5 Three-Layer Fuzzy Support Vector Regression-Based Intention Understanding Model 141
9.6 Experiments 142
9.6.1 Experiment Setting 143
9.6.2 Experiments on Three-Layer Fuzzy Support Vector Regression Based Intention Understanding Model 143
9.7 Summary 156
References 157

10 Dynamic Emotion Understanding Based on Two-Layer Fuzzy Fuzzy Support Vector Regression-Takagi-Sugeno Model 161
10.1 Introduction 161
10.2 Dynamic Emotion Recognition Using Candide3-Based Feature Point Matching 164
10.3 Two-Layer Fuzzy Support Vector Regression for Emotional Intention Understanding 166
10.4 Two-Layer Fuzzy Support Vector Regression Takagi-Sugeno Model for Emotional Intention Understanding 169
10.5 Experiments 171
10.5.1 Experimental Environment 171
10.5.2 Self-Built Data 171
10.5.3 Experiments on Dynamic Emotion Recognition and Understanding 173
10.6 Summary 180
References 180

11 Emotion-Age-Gender-Nationality Based Intention Understanding Using Two-Layer Fuzzy Support Vector Regression 183
11.1 Introduction 183
11.2 Two-Layer Fuzzy Support Vector Regression 185
11.2.1 Support Vector Regression 185
11.2.2 Two-Layer Fuzzy Support Vector Regression 187
11.3 Intention Understanding 189
11.3.1 Emotion Based Intention Understanding 189
11.3.2 Characteristics Analysis 190

11.4 Intention Understanding Model 197
11.4.1 Emotion Recognition 198
11.4.2 Questionnaire 199
11.4.3 ID Mapping 199
11.5 Two-Layer Fuzzy Support Vector Regression Based Intention Understanding 200
11.5.1 Intention Generation by Fuzzy Inference 201
11.6 Memory Retrieval for Intention Understanding 201
11.7 Experiments 202
11.7.1 Experiment Setting 202
11.7.2 Experiments on Two-Layer Fuzzy Support Vector Regression Based Intention Understanding Model 204
11.8 Summary 210
References 212

12 Emotional Human-Robot Interaction Systems 215
12.1 Introduction 215
12.2 Basic Emotional Human-Robot Interaction Systems 218
12.3 Design of Emotional Human-Robot Interaction System 220
12.4 Summary 221
References 221

13 Experiments and Applications of Emotional Human-Robot Interaction Systems 223
13.1 Introduction 223
13.2 Emotional Interaction Scenario Setting 225
13.3 Multi-modal Emotion Recognition Experiments Based on Facial Expression, Speech and Gesture 227
13.4 Emotional Intention Understanding Experiments Based on Emotion, Age, Gender, and Nationality 232
13.5 Application of Multi-modal Emotional Intention Understanding 238
13.5.1 Self-built Data 238
13.5.2 Experiments on Dynamic Emotion Recognition and Understanding 239
13.6 Summary 243
References 243

Index 245

Chapter 1
Introduction

Human-robot interaction technology has gradually shifted from computer-centered to human-centered, as natural human-robot interaction has become a new direction of human-robot interaction technology development. Emotional interaction is the core and foundation of natural human-robot interaction, which makes robots capable of emotional communication and more humanized by processing the emotional information [1] in the process of human-robot interaction. The emergence of big data, the innovation of algorithms, the improvement of computer computing abilities, and the evolution of network facilities are driving the development of artificial intelligence. Intelligentization has become an important direction for the development of technology and industry [2]. Now, the robot revolution has entered the era of "Internet + artificial intelligence + emotion", which requires robots to have the ability of emotional cognition [3].

Intelligent emotional robots increasingly appear in people's life. Their system construction is usually based on the ordinary humanoid robot, emotional information retrieval framework, emotional understanding framework, emotional robot interaction framework, eventually form a shape similar to human, inner emotional robot system with human emotions, cognitive ability [4]. For example, Buddy, the first emotional companion family robot launched by Blue Frog Robotics, can naturally express emotions following its interactions with family members [5]. Humanoid robot Sophia, developed by Hanson robotics, is the first robot in history to obtain citizenship. It can display more than 62 facial expressions, understand language and remember interactions with humans [6]. Erica, an intelligent robot developed by professor Hiroshi Ishiguro's team at Osaka university in Japan and a research team at Kyoto university, can talk with people fluently, with sounds and expressions very similar to those of human beings. In addition, Erica once reported the news as a news anchor on a Japanese TV station in 2018 [7].

Existing emotional robot system mainly includes two aspects, a study is to let the robot to identify human emotional state and based on the interaction of feedback to adjust its behavior, this kind of research by identifying human facial expressions,

L. Chen et al., *Emotion Recognition and Understanding for Emotional Human-Robot Interaction Systems*, Studies in Computational Intelligence 926,
https://doi.org/10.1007/978-3-030-61577-2_1

voice signal, body posture and physiological signals for emotional information, and then according to the acquired user emotional response behavior guide robot, realizing human-robot interaction [8]. For example, the literature proposed a companion robot system with the ability of emotional interaction. The robot uses visual and audio sensors to detect the user's basic emotions. The robot plays appropriate music according to the user's emotional state, and generates a soothing odor [9]. The robot system can also automatically navigate and follow the user's side to accompany the user [10]. Literature reports on a robot emotion recognition and rehabilitation system based on browser interface, which integrates physiological signals and facial expressions to identify users' emotional states, and then combines the technology of emotional expression to alleviate users' negative emotions or enhance their positive emotions [11].

Another kind of research explores the multi-modal emotional expression of robots [12]. By adjusting the facial expression, speech synthesis mechanism, posture and movements of humanoid robots, the robots are endowed with as rich emotional expression capacity as human beings [13]. The humanoid robot ROBIN was developed in literature, which can express almost any facial expression of human beings and generate expressive gestures and body gestures. At the same time, it is also equipped with a voice synthesizer to realize the transformation, selection, expression, stimulation and evaluation of emotions, as well as some auxiliary functions [14]. Literature has developed an emotional robot system to accompany autistic children. By playing games and communicating with autistic children, emotional robots can identify the emotions expressed by interactors and correctly classify them [15].

At the same time, research institutions are paying more and more attention to the research of affective computing and intelligent interaction [16]. It is estimated that robots will rapidly penetrate into the service industry from the industrial field, showing a broader market space than industrial robots [17]. However, it is very important for service robots to improve their service quality to integrate emotional factors into service robots and have emotional cognitive ability, so that they can understand human needs and realize a virtuous circle of emotional communication and satisfaction of needs.

Facing the urgent needs of human-robot emotional interaction system and emotional robot in domestic and foreign markets, the research of emotional robot system will surely show a new direction for the future application of natural interaction. The ultimate purpose of constructing the emotional robot system is to make the robot exhibit appropriate behavioral response based on the understanding of the emotional state of the communicating object, so as to adapt to the constant changes of the user's emotion, and thus optimize the interactive service. Even more anthropomorphic robot, however, is also the basic, no general intelligent, system can close to the human level, the existing emotional robot is not short of human emotional ability, how to let the machine have more accurate emotion recognition and the intention of the deeper understanding ability, and guide the act more natural, proper response, is essential to the natural human computer interaction.

At present, some progress has been made in the research of emotional robot system related to the understanding of emotional intention and natural interaction technology. However, the research is still in its infancy, and there are still many problems to be further solved, mainly reflected in the following aspects [18]: The research on facial expression, speech emotion, body posture, hand gestures and body language), physiological signals (EEG) and pulse on development of recognition, but at this stage of multimodal information fusion method is not sufficiently considering the modal characteristics in identifying different emotion plays different role, under the condition of not both the correlation and differences between different modal characteristics [19]; The research on the understanding of emotional intention is still in the preliminary stage, and the deep structure of emotional intention perception is relatively simple, which has not yet realized the multi-dimensional understanding of emotional intention [20]. At the present stage, emotional models are not closely related to human-robot interaction means, so it is particularly crucial to study more general and fuzzy emotional models that conform to human emotional changes and introduce them into the behavioral decision-making and coordination mechanism of robots [21]. There are still many deficiencies in the emotional robot system, such as not considering the environmental information and scene information, which will also affect the emotional expression of the interactor, or not considering the deep intention information behind the user's emotional expression, and lack of deep cognitive analysis of the user's behavior, etc. [22], which cannot form a deep social relationship [23].

Therefore, on the basis of the existing emotional robot system considering modal emotional information and environmental information, introducing emotional intention understanding framework, combined with the geometrical visual SLAM algorithmto build a robot autonomous positioning and visual navigation model, realize the emotional robot in emotional interaction in the process of environmental perception, so as to take advantage of new human-robot interaction technology, build emotional robot system with emotional intention interaction ability, is particularly necessary.

In order to realize a human-robot interaction system with certain emotion recognition and intention understanding, and to establish a natural and harmonious human-robot interaction process, this section proposes a human-robot interaction system scheme based on affective computing. First, the overall architecture design and application scenario of the system are introduced, and then the application experiment based on the emotional human-robot interaction system is introduced.

1.1 Emotion Feature Extraction and Recognition

In our daily communication, the human being able to capture the emotional change of the other person by observing the facial expression and the body gesture, listening to the voice. This is because the human brain is capable of sensing and understanding the information that reflects a person's emotional state in his voice and the visual signal (such as the special speech word, the change of the intonation and facial expression, etc.). The multi-modal emotion recognition is the simulation of the human emotion perception and understanding process by robots, and the task of which is to extract the emotion feature from multi-modal signals, and find out the mapping relationship between the emotional features and human emotions.

Facial expression, speech signal and body gesture often appear in the process of human-robot interaction at the same time, which are used to analyze and inference each other's emotional states and intention in real time, and then guide them to make different reactions [24]. It can be imagined that if robots can be given the same visual, auditory, and cognitive abilities as humans, they could be able to make personalized and adaptive responses to each other's states, just like humans. In the aspect of emotion recognition, most of the studies are developed based on the different modes emotional data. The multi-modal emotion recognition by constructing a corresponding feature set based on different mode of emotion information can effectively reflect the emotional state.

The current research on affective computing mainly focuses on facial expression recognition, speech emotion recognition and body gesture recognition. In the emotional human-robot interaction, multi-modal emotion recognition is mainly divided into two steps: multi-modal emotion feature extraction and emotion recognition, in which the multi-modal emotion feature extraction is a very important link. The ability of feature sets to represent emotional information directly affects the results of emotion recognition.

Facial expression is the most intuitive way to express emotions. At present, there are two mainly types of methods for extracting facial features: deformation-based facial feature extraction and motion-based facial feature extraction. The method of extracting facial features based on deformation mainly include principal component analysis (PCA), linear discriminant analysis (LDA), geometric features and models, Gabor wavelet transform, etc. The method of extracting facial features based on motion is to treat the facial expression as wording, and analyze and recognize facial expressions through the information of facial motion changes. The main core is to use motion changes as recognition features. At present, the motion feature extraction methods mainly include optical flow method and feature point tracking method.

In speech emotion recognition, speech emotion features can be divided into acoustic and language features [25]. The two types of feature extraction methods and their contribution to speech emotion recognition are different depending on the selected speech database. If the speech database is a text-based database, the language features can be ignored and if the database is close to current real life, language features will play a great role. The acoustic features used for speech emotion recognition can be summarized into prosodic features, spectral-based features, and voice quality features. These features are often extracted in frames, and participate in emotion recognition in the form of global feature statistics. Some studies suggest that various gestures and movements expressed by the body movements in interactive behaviors can express or assist in articulating people's thoughts, emotions, intentions, etc. So that body gesture is important for understanding emotional communication. Feature extraction of body gesture mainly includes global feature extraction and local feature extraction. The global features include color, texture, motion energy image, and motion history image, etc. The local features include gradient histogram, Scale-invariant feature transform (SIFT), speed-up robust features (SURF), spatio-temporal points of interest, etc.

Most of the current emotion databases describe emotion as an adjective label form (such as anger, happiness, sadness, fear, surprise and neutrality), which is a discrete emotion description model. Therefore, it's always attributed to the standard pattern classification problem for emotion recognition based on the discrete emotion description model and the corresponding emotion database. When the training data and test data come from different people or different databases, in addition to requiring good emotional representation ability for emotional features, the design of the emotion recognition classifier also has higher requirements. In what follows we mainly introduce the methods of discrete emotion recognition [26].

The research on single-modal emotion recognition algorithm is relatively mature [27–29]. However, although multi-modal emotion recognition has received widespread attention, there are few related studies. Most of them are emotion recognition based on bi-modal information, such as speech-facial emotion recognition [30], posture-facial emotion recognition [31], physiological signal-facial [32] emotion recognition, etc. Multi-modal information fusion method is very important to make full use of multi-modal emotion information and improve the performance of emotion recognition. Information fusion, as the theoretical basis of multimodal emotion recognition, covers a wide range of fields. When performing multi-modal emotional interactions, multi-channel sensors are used to obtain the interviewer's signals of different emotional state, and then data fusion and decision-making are performed, the key of which is the multi-modal emotion recognition algorithm. The emotional feature data of each channel are fused, and the decision is made according to certain rules, so as to determine the emotion category attributes corresponding to the multi-modal information. Multi-modal emotional information fusion can be divided into feature-level fusion and decision-level fusion. The overall framework is shown in Fig. 1.1.

Feature-level fusion first extracts features from the original information obtained by the sensors, and then comprehensively analyzes and processes the information [33]. In general, the extracted feature information should be a sufficient statistics of the pixel information, and the multi-sensor data should be classified, aggregated, and integrated according to the feature information. The fusion system first preprocesses the data to complete data calibration, and then implements parameter correlation and state vector estimation. In multi-modal information fusion, the feature-level fusion strategy is to first extract the emotional feature data in each mode separately, and then cascade the feature data of all modalities into a feature vector for emotion recognition. Only one classifier is designed for the emotional feature data, and the output of the classifier is the predicted result of emotional type. Figure 1.2 shows the feature level fusion of multi-modal emotion.

As for decision-level fusion, before proceeding with the fusion, the corresponding processing components of each local sensor have independently completed the decision-making or classification task [34]. The essence is to coordinate according to certain criteria and the credibility of each sensor to make a global optimal decision. Decision level fusion is a joint decision result, which is theoretically more precise

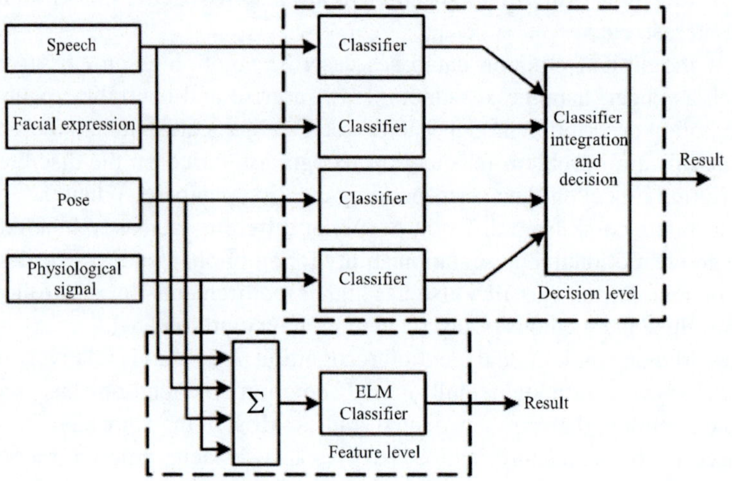

Fig. 1.1 The overall framework

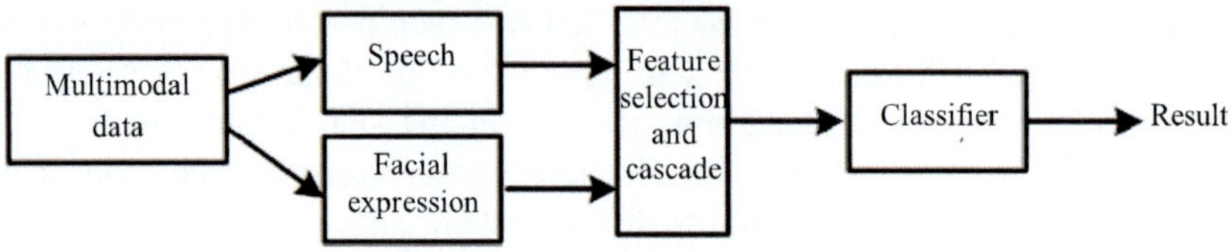

Fig. 1.2 The feature level fusion of multi-modal emotion

and clear than any single sensor decision [35]. At the same time, it is also a high-level fusion. The result can provide a basis for the final decision. Therefore, decision-level fusion must fully consider the needs of specific decision-making issues, and make full use of the various types of feature information of the measurement object, then use appropriate fusion technology. Decision-level fusion is directly aimed at specific decision-making goals, and the result of which directly affects the level of decision-making. In the emotion recognition, the decision-level fusion strategy first designs a corresponding emotion classifier for the emotional feature data, and then makes a decision on the output of each classifier to synthesize the final emotion recognition result according to certain decision rules [36]. Figure 1.3 shows the decision-level fusion of multi-modal emotional feature data. The methods used for decision-level fusion include Bayesian reasoning, Dempster-Shafer evidence theory, and fuzzy reasoning, etc.

The current emotion recognition methods mainly include neural network, support vector machine, and extreme learning machine (ELM) [37]. Neural network algorithms are widely used in the field of emotion recognition. Back propagation neural

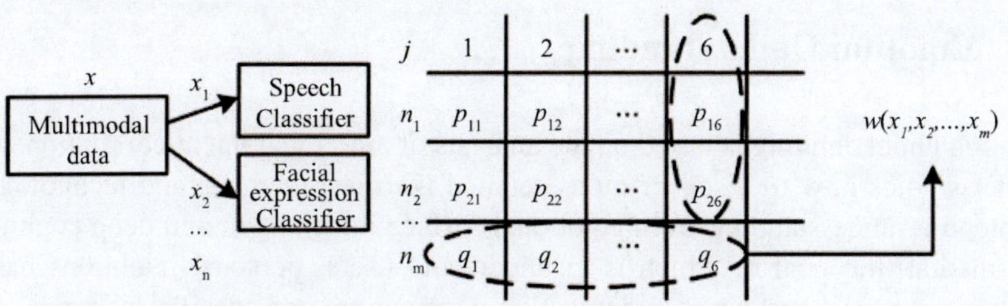

Fig. 1.3 The decision-level fusion of multi-modal emotional feature data

networks (BPNN) are characterized by forward signal transfer and error back propagation, which also make the calculation process structured into two major steps. During the forward transfer of the signal, the signal enters the hidden layer from the input layer, and the result is formed in the output layer. The processing result of the previous layer will directly affect the processing effect of the subsequent layer. If the output layer fails to achieve the desired result, it will switch to back propagation, adjust the network weights and thresholds according to the prediction error until the output result is close to the desired output result. Deep learning algorithms have also been applied to the field of emotion recognition for learning and recognition of emotion features. Deep learning combines low-level features to form more abstract high-level representation or features to discover distributed feature representations of data. At present, typical deep learning models include: Convolution Neural Network (CNN), Deep Belief Network (DBN), Autoencoder, and Recurrent Neural Network (RNN) [38, 39].

Support vector machine (SVM) has outstanding advantages in processing small samples and cope with non-linear problems. The essence of SVM is to find the optimal linear classification hyperplane, which includes two cases with sample linearly separable and sample nonlinearly separable. For the first case, SVM tries to find the best among the classification lines that completely separate the samples. For the second case, SVM uses the kernel function to solve the linear discrimination in the high-dimensional feature space, and also solves the problem of large amount of computation in high-dimensional feature space. Emotional feature parameters are not completely linearly separable in the input space, so non-linearly separable cases are used for emotion recognition.

The extreme learning machine (ELM) has only a single hidden layer. Unlike traditional learning theory, which needs to adjust all parameters of the feedforward neural network [40], it randomly assigns input weights and thresholds to the neurons. Then the weights are calculated and output through the regularization principle, so that the neural network can still approximate continuous system. Since it is proved that the random acquisition of the hidden layer node parameters of the single hidden layer neural network will not affect the network's convergence ability, the training of the over-limit learning machine is much shorter than the traditional BP neural network and SVM.

1.2 Emotion Understanding

Emotion understanding is based on the analysis of emotional intent correlation, and further studies how to discover more detailed theories, methods and technologies of intention understanding from emotional surface information and deep cognitive information, the goal of which is to understand users' personal intentions based on their emotions, surface communication information, and specific scenarios, and ultimately achieve natural harmony of human-robot interaction [41].

Humans are inherently capable of expressing, predicting, and understanding intentions, regardless of whether the intentions are expressed in an explicit or implicit manner [42]. Intent is an important part of interpersonal communication and the human cognitive system. Although it cannot be directly obtained, it can be inferred from the behavior, physiological indicators, and atmosphere. The research on intent understanding mainly focused on the research of physical and psychological behavior. With the continuous development of artificial intelligence and computer technology, many scholars have begun to introduce intention understanding into human-robot interaction in an attempt to make machines have the same intention understanding ability as humans, thereby promoting harmonious human-robot interaction.

The research on intent understanding mainly focused on the research of physical and psychological behavior. With the continuous development of artificial intelligence and computer technology, many scholars have begun to introduce intention understanding into human-robot interaction in an attempt to make machines have the same intention understanding ability as humans, thereby promoting harmonious human-robot interaction. In order to realize the emotional intention understanding, most researchers mainly realize the understanding of emotional intention by studying the mapping relationship between emotion and intention. In our previous research [43], in order to deeply understand human internal thinking, an intention understanding model based on two-layer fuzzy support vector regression is proposed in human-robot interaction, where fuzzy c-means clustering is used to classify the input data, and intention understanding is mainly obtained by emotion, with identification information such as age, gender, and nationality. It aims to realize the transparent communication by understanding customers order intentions at a bar, in such a way that the social relationship between bar staffs and customers becomes smooth. To demonstrate the aptness of intention understanding model, experiments are designed in term of relationship between emotion-age-gender-nationality and order intention. Chen et al. [44] proposed a Two-layer Fuzzy SVR-TS Model for emotion understanding in human-robot interaction, where the real-time dynamic emotion recognition is realized by using Candide3 based feature point matching method, and emotional intention understanding is obtained mainly based on human emotions and identification information. It aims to make robots capable of recognizing and understanding human emotions, in such a way that make human-robot interaction run smoothly. Chen et al. [45] proposed a three-layer weighted fuzzy support vector regression (TLWFSVR) model for understanding human intention, and it is based on the emotion-identification information in human-robot interaction. The TLWFSVR model consists of three layers, including adjusted weighted kernel fuzzy c-means for data clustering, fuzzy support vector regressions (FSVR) for information understanding, and weighted fusion for intention understanding. It aims to guarantee the quick convergence and satisfactory performance of the local FSVR via adjusting the weights of each feature in each cluster, in such a way that importance of different emotion-identification information is represented. Moreover, smooth human-oriented interaction can be obtained by endowing robot with human intention understanding capabilities.

1.3 Emotional Human-Robot Interaction System

With the integration and rapid development of robotics and artificial intelligence, more and more robots have begun to enter people's daily life. The personification of robots will definitely be an important direction for future development. Researchers have tried to give robots a human form to make them more acceptable to us, so there have been many humanoid robots that can execute human commands, which performs in a superb way in tour guide welcome, education and teaching, disaster relief, and rehabilitation [23, 46].

However, people's expectations for humanoid robots don't stop there. Robots can't meet the needs of humans by carrying out some mechanical repetitive commands. They want robots to posses human-like intelligence, so that robots are endowed with the ability of emotion perception and expression. It is an important direction for future development to build a natural and autonomous cognitive robot interaction system that meets the psychological needs. The construction of the emotional robot system is generally based on the ordinary humanoid robot, constructing emotional information acquisition frame, emotional understanding frame, and emotional robot interaction frame, which finally form an emotional robot system that resembles a humanoid in shape and has human emotional cognitive capabilities. Existing research on emotional robot systems mainly includes two aspects, one of which realizing emotional information acquisition by recognizing facial expressions, speech signals, body gestures, etc., and then guide the robot to make behavioral responses based on the acquired emotions. For example, [47] proposed a companion robot system with emotional interaction capabilities. The robot detects the user's basic emotions through visual and audio sensors. According to the user's emotional state, the robot will play appropriate music, and generate a scent that soothes the user's mood. The robotic system can also automatically navigate and accompany the user beside him. Somchanok and Michiko [13] developed a robot-based emotion recognition and rehabilitation system based on a browser interface, which combined physiological signals and facial expressions to identify the user's emotional state, and then combined emotional expression technology to alleviate the negative emotions of the user or strengthen the positive emotions. Another research explores the robot's multi-modal emotional expression. By adjusting the facial expressions, speech mechanisms, and body gestures, robots are endowed with rich emotional expression capabilities like humans. Klug and Zell [48] developed a humanoid robot ROBIN, which can express almost any facial expression of human beings, and generate expressive gestures. It is also equipped with a speech synthesizer to achieve emotional transformation, selection, expression, and evaluate, and some auxiliary functions. Boccanfuso et al. [49] developed an emotional robot system that accompanies children with autism. By communicating with children with autism, the emotional robot can recognize the emotions expressed by the children and correctly classify them.

There are many shortcomings in current emotional robot systems. For example, the current research does not consider environmental and scene information, which will affect the emotional expression of the operator. At the same time, it does not take into account the deep intention information behind the emotional expression, and lacks a cognitive analysis of user behavior. It is necessary to comprehensively consider multi-modal emotional information and environmental information on the basis of existing emotional robot systems, introduce an emotional intention understanding framework, and build an emotional robot system with emotional intention interaction capabilities under a networked architecture.

1.4 Organization of This Book

This book first introduces the essential knowledge of multi-modal emotion recognition, emotional intent understanding and emotional human-robot interaction system in the introduction, and theoretically explains the complete process of emotional human-robot interaction. Then combined with specific algorithms, we briefly describe some of our research results, including algorithm theory of multi-modal emotion feature extraction, multi-modal emotion recognition, and emotion intention understanding. Finally, some simulation experiments and application results of our emotional human-robot interaction system are reported.

This book is structured as follows. In Chap. 2, combined with the characteristics of facial expression, speech, and gesture, the construction method of multimodal emotional feature set is systematically described. In Chap. 3, Softmax regression-based deep sparse autoencoder network is proposed to recognize facial emotion. Chapter 4 introduces AdaBoost-KNN using adaptive feature selection with direct optimization is proposed for dynamic emotion recognition. In Chap. 5, the weight-adapted convolution neural network is proposed to extract discriminative expression representations for recognizing facial expression. Chapter 6 presents the two-layer fuzzy multiple random forest for speech emotion recognition. In Chap. 7, the two-stage fuzzy fusion based-convolution neural network is presented in this chapter for dynamic emotion recognition by using both facial expression and speech modalities. Chapter 8 presents the Dempster-Shafer theory based on multi-SVM to deal with multimodal gesture images for intention understanding. In Chap. 9, the three-layer weighted fuzzy support vector regression model is proposed for understanding human intention, and it is based on the emotion-identification information in human-robot interaction. Chapter 10 proposes Two-layer fuzzy support vector regression-Takagi–

Sugeno (TLFSVR-TS) model for emotion understanding. In Chap. 11, an intention understanding model based on two-layer fuzzy support vector regression is proposed in human-robot interaction. Chapter 12 introduces the basic construction method of emotional human-robot interaction systems based on multi-modal emotion recognition and emotion intention understanding. In Chap. 13, some simulation experiments and application results of our emotional human-robot interaction system are shown.

References

1. D. Schuller, B.W. Schuller, The age of artificial emotional intelligence. Computer **51**(9), 38–46 (2018)
2. F. Li, J. Feng, M. Fu, A natural human-robot interaction method in virtual roaming, in *Proceedings of the 2019 15th International Conference on Computational Intelligence and Security (CIS)* (2019)
3. A. Vinciarelli, M. Pantic, H. Boulard, Social signal processing: survey of an emerging domain. Image Vis. Comput. **27**(12), 1743–1759 (2009)
4. J.C. Gómez, A.G. Serrano, P. Martínez, Intentional processing as a key for rational behaviour through natural interaction. Interact. Comput. **18**(6), 1419–1446 (2006)
5. J. Fan, M. Campbell, B. Kingsbury, Artificial intelligence research at IBM. IBM J. Res. Dev. **55**(5), 16: 1–16: 4 (2011)
6. X.Q. Zheng, M. Shiomi, T. Minato, H. Ishiguro, What kinds of robot's touch will match expressed emotions. IEEE Robot. Autom. Lett. **5**(1), 127–134 (2019)
7. P. Christopher, L. Johnson, D.A. Carnegie, Mobile robot navigation modulated by artificial emotions. IEEE Trans. Syst. Man Cybern. Part B (Cybern.) **40**(2), 469–480 (2009)
8. Blue Frog Robotics, Buddy: the first emotional companion robot (2018). http://www.Bluefrogrobotics.com/en/buddy_the-emotionnal-robot/
9. Hanson Robotics, Sophia (2016). http://www.hansonrobotics.com/robot/sophia/
10. Live Science, Meet Erica, Japan's next robot news anchor (2018). https://www.livescience.com/61575-erica-robot-replace-japanese-news-anchor.html
11. S.B. Lee, S.H. Yoo, Design of the companion robot interaction for supporting major tasks of the elderly, in *Proceedings of the 14th International Conference on Ubiquitous Robots and Ambient Intelligence (URAI)* (2017)
12. J.H. Lui, H. Samani, K.Y. Tien, An affective mood booster robot based on emotional processing unit, in *Proceedings of 2017 International Automatic Control Conference*, pp. 1–6 (2018)
13. T. Somchanok, O. Michiko, Emotion recognition using ECG signals with local pattern description methods. Int. J. Affect. Eng. **15**(2), 51–61 (2016)
14. S. Saunderson, G. Nejat, It would make me happy if you used my guess: comparing robot persuasive strategies in social human-robot interaction. IEEE Robot. Autom. Lett. **4**(2), 1707–1714 (2019)
15. M. Klug, A. Zell, Emotion-based human-robot interaction, in *Proceedings of IEEE International Conference on Computational Cybernetics*, pp. 365–368 (2013)
16. L. Boccanfuso, E. Barney, C. Foster, Emotional robot to examine differences in play patterns and affective response of children with and without ASD, in *Proceedings of ACM/IEEE International Conference on Human-Robot Interaction*, pp. 19–26 (2016)
17. M. Ficocelli, J. Terao, G. Nejat, Promoting interactions between humans and robots using robotic emotional behavior. IEEE Trans. Cybern. **46**(12), 2911–2923 (2016)
18. Y. Ding, X. Hu, Z.Y. Xia, Y.J. Liu, D. Zhang, Inter-brain EEG feature extraction and analysis for continuous implicit emotion tagging during video watching. IEEE Trans. Affect. Comput. https://doi.org/10.1109/TAFFC.2018.2849758

19. J.C. Peng, Z. Lina, Research of wave filter method of human sphygmus signal. Laser Technol. **40**(1), 42–46 (2016)
20. C.L.P. Chen, Z. Liu, Broad learning system: an effective and efficient incremental learning system without the need for deep architecture. IEEE Trans. Neural Netw. Learn. Syst. **29**(1), 10–24 (2018)
21. Z.T. Liu, F.Y. Dong, K. Hirota, M. Wu, D.Y. Li, Y. Yamazaki, Emotional states based 3-D fuzzy atmosfield for casual communication between humans and robots, in *Proceedings of International Conference on Fuzzy Systems*, Taipei, pp. 777–782 (2011)
22. S.Y. Jiang, C.Y. Lin, K.T. Huang, Shared control design of a walking-assistant robot. IEEE Trans. Control Syst. Technol. **25**(6), 2143–2150 (2017)
23. X. Lu, W. Bao, S. Wang, Three-dimensional interfacial stress decoupling method for rehabilitation therapy robot. IEEE Trans. Ind. Electron. **64**(5), 3970–3977 (2017)
24. A. Mehrabian, Communication without words. Psychol. Today **2**(4), 53–56 (1968)
25. B. Schuller, A. Batliner, S. Steidl, D. Seppi, Recognising realistic emotions and affect in speech: state of the art and lessons learnt from the first challenge. Speech Commun. **53**(9), 1062–1087 (2011)
26. L. Chen, S.K. Zheng, Speech emotion recognition: features and classification models. Digit. Signal Process. **22**(6), 1154–1160 (2012)
27. L.F. Chen, M.T. Zhou, W.J. Su, M. Wu, J.H. She, K. Hirota, Softmax regression based deep sparse autoencoder network for facial emotion recognition in human-robot interaction. Inf. Sci. **428**, 49–61 (2018)
28. L.F. Chen, Y. Feng, M.A. Maram, Y.W. Wang, M. Wu, K. Hirota, W. Pedrycz, Multi-SVM based Dempster-Shafer theory for gesture intention understanding using sparse coding feature. Appl. Soft Comput. https://doi.org/10.1016/j.asoc.2019.105787
29. L.F. Chen, W.J. Su, Y. Feng, M. Wu, J.H. She, K. Hirota, Two-layer fuzzy multiple random forest for speech emotion recognition in human-robot interaction. Inf. Sci. **509**, 150–163 (2020)
30. Y. Kim, E.M. Provost, ISLA: temporal segmentation and labeling for audio-visual emotion recognition. IEEE Trans. Affect. Comput. https://doi.org/10.1109/TAFFC.2017.2702653
31. P. Barros, D. Jirak, C. Weber, Multimodal emotional state recognition using sequence-dependent deep hierarchical features. Neural Netw. **72**, 140–151 (2015)
32. M. Soleymani, S. Asghariesfeden, Y. Fu, Analysis of EEG signals and facial expressions for continuous emotion detection. IEEE Trans. Affect. Comput. **7**(1), 17–28 (2016)
33. Y. Xie, R. Liang, Z.L. Liang, C.W. Huang, C.R. Zou, Speech emotion classification using attention-based LSTM. **27**(11), 1675–1685 (2019)
34. C.H. Wu, W.L. Wei, J.C. Lin, W.Y. Lee, Speaking effect removal on emotion recognition from facial expressions based on eigenface conversion. IEEE Trans. Multimed. **15**(8), 1732–1744 (2013)
35. C.H. Wu, W.B. Liang, Emotion recognition of affective speech based on multiple classifiers using acoustic-prosodic information and semantic labels. IEEE Trans. Affect. Comput. **2**(1), 10–21 (2011)
36. A. Hariharan, M.T.P. Adam, Blended emotion detection for decision support. IEEE Trans. Hum.-Mach. Syst. **45**(4), 510–517 (2015)
37. Z.T. Liu, M. Wu, W.H. Cao, L.F. Chen, J.P. Xu, R. Zhang, M.T. Zhou, J.W. Mao, A facial expression emotion recognition based human-robot interaction system. IEEE/CAA J. Autom. Sin. **4**(4), 668–676 (2017)
38. Y. Li, J.B. Zeng, S.G. Shan, X.L. Chen, Occlusion aware facial expression recognition using CNN with attention mechanism. IEEE Trans. Image Process. **28**(5), 2439–2450 (2019)
39. T. Zhang, W.M. Zheng, Z. Cui, Y. Zong, Y. Li, Spatial-Temporal Recurrent Neural Network for Emotion Recognition. IEEE Trans. Cybern. **49**(3), 839–847 (2019)
40. L.A. Bugnon, R.A. Calvo, D.H. Milone, Dimensional affect recognition from HRV: an approach based on supervised SOM and ELM. IEEE Trans. Affect. Comput. **11**(1), 32–44 (2020)
41. M. Wu, Z.T. Liu, L.F. Chen, *Affective Computing and Emotional Robot System* (Beijing Science Press, Peking, 2018)
42. D.M. Wegner, *The Illusion of Conscious Will* (MIT Press, Cambridge, 2002)

43. L.F. Chen, Z.T. Liu, M. Wu, M. Ding, F.Y. Dong, K. Hirota, Emotion-age-gender-nationality based intention understanding in human-robot interaction using two-layer fuzzy support vector regression. Int. J. Soc. Robot. **7**(5), 709–729 (2015)
44. L.F. Chen, M. Wu, M.T. Zhou, Dynamic emotion understanding in human-robot interaction based on two-layer fuzzy SVR-TS model. IEEE Trans. Syst. Man Cybern. Syst. PP(99), 1–12 (2017)
45. L.F. Chen, M.T. Zhou, M. Wu, Three-layer weighted fuzzy support vector regression for emotional intention understanding in human-robot interaction. IEEE Trans. Fuzzy Syst. PP(99), 1 (2018)
46. J. Turner, Q. Meng, G. Schaefer, Distributed task rescheduling with time constraints for the optimization of total task allocations in a multi-robot system. IEEE Trans. Cybern. **48**(9), 2583–2597 (2018)
47. J.H. Lui, H. Samani, K.Y. Tien, An affective mood booster robot based on emotional processing unit, in *Proceedings of 2017 International Automatic Control Conference*, Pingtung, Taiwan, China, pp. 1–6 (2018)
48. M. Klug, A. Zell, Emotion-based human-robot interaction, in *Proceedings of the 9th IEEE International Conference on Computational Cybernetics*, Tihany, Hungary, pp. 365–368 (2013)
49. L. Boccanfuso, E. Barney, C. Foster, Emotional robot to examine differences in play patterns and affective response of children with and without ASD, in *Proceedings of the 11th ACM/IEEE International Conference on Human-Robot Interaction*, Christchurch, New Zealand, pp. 19–26 (2016)

Chapter 2
Multi-modal Emotion Feature Extraction

Multi-modal emotion feature extraction is an indispensable part of multi-modal emotion recognition. In order to make effective use of emotion information in multi-modal emotion recognition, the corresponding feature extraction method should be adopted according to the characteristics of emotion information of different modes. In this chapter, three methods are proposed for three modes' features, namely, regions of interest based feature extraction in facial expression, sparse coding-SURF based feature extraction in body gesture and FCM based feature extraction in the speech emotion.

2.1 Introduction

In emotion recognition with multi-modal information, the emotional feature extraction and multi-modal fusion are the most important steps. In multi-modal emotion recognition, the abilities of feature set to represent emotion information and its overall performance directly affect the result of emotion recognition and the overall interaction effect. In the construction of multi-modal emotional feature set, different feature extraction channels are generally selected to extract corresponding information of different modes in parallel. This section introduces the extraction methods of the emotional features of each mode based on the characteristics of the three modes of facial, gesture and speech emotion.

Facial emotion feature extraction is the most important step in facial emotion recognition. According to the type of input image, the emotion feature extraction algorithm can be divided into the feature extraction algorithm based on static image and the feature extraction algorithm based on dynamic image sequence. The feature extraction algorithm based on static image only considers the spatial information of single frame image and the geometric structure information of face. In this feature extraction algorithm, geometric feature extraction algorithm and appearance feature

L. Chen et al., *Emotion Recognition and Understanding for Emotional Human-Robot Interaction Systems*, Studies in Computational Intelligence 926,
https://doi.org/10.1007/978-3-030-61577-2_2

extraction algorithm are commonly used. Typical geometric feature-based extraction algorithms mainly include Active Shape Models (ASM) [1], Active Appearance Models (AAM) [2] and Facial Action Coding System (FACS) [3]. Appearance feature based extraction algorithm aims to use the whole face or specific area in the face image to reflect the basic information of the face image, especially the subtle changes of the face. A lot of research is based on ASM and AAM models to extract the facial feature points location [4] to construct the facial geometric features. Based on the characteristics of both FACS extraction such as [5–7]. In addition, there are some geometric feature extraction methods, such as the SIFT method and the distance between facial horizontal curve [8, 9]. Based on appearance feature extraction algorithm is designed to use the whole face or specific areas from the face image to reflect the basic information of the face image, especially the changes in my face. At present, Local Binary Pattern (LBP) [10] and Gabor wavelet transform are the main algorithms for appearance feature extraction [11]. Some studies have reduced the dimension of Gabor features [12]. On the original LBP operator, some researches have realized the expansion of time dimension [13].

Feature extraction algorithm based on dynamic image sequence needs to consider the motion information of emotion image. Facial muscle movement and facial deformation have become the contents to be extracted by feature extraction method based on dynamic image sequence. Common algorithms include optical flow method [14], feature point tracking method [15] and difference image method. Researches on motion features of facial expressions extracted based on optical flow method, such as [16, 17]. Methods based on feature point tracking are often combined with AMM [18–20]. Deep learning can be used for both static image feature extraction and dynamic sequence, such as 3D convolutional network [21] and enhanced depth confidence network [22].

The categories of phonetic emotional features can be roughly divided into acoustic features and linguistic features [23]. The two kinds of feature extraction methods and their contributions to speech emotion recognition are also quite different depending on the selected corpus. If the selected corpus is a text-based database, the language features can be ignored. If the selected corpus is the real corpus close to the real life, the language features will play a great role. Most previous scholars focused on the study of acoustic characteristics. The acoustic features used for speech emotion recognition can be roughly summarized into three types: prosodic features [24], spectral related features and sound quality features [25]. These features are often extracted in frames, but are involved in emotion recognition in the form of global feature statistics. The unit of global statistics is generally an auditory independent statement or word, and the commonly used statistical indicators are extremum, extremum range, variance, etc.

In addition to facial emotion and phonological emotion, gesture is an indispensable means of emotion in human communication. Wallbott [26] analyzed the emotional content of gesture gestures with coding schema, indicating that gesture gestures can represent specific emotions. DeMeijer [27] also demonstrated that different emotional categories can be inferred from the intensity and type of physical activity. Feature extraction methods of attitude actions mainly include global feature extraction method and local feature extraction method [28]. Among the feature extraction methods based on global features, Motion Energy Image (MEI) and Motion History Image (MHI) are commonly used [29], and feature extraction based on optical flow. Feature extraction methods based on local features are implemented in a bottom-up manner, and can be divided into feature extraction based on points of interest [30] and feature extraction based on tracks [31].

In this chapter, three methods are proposed for three modes' features, namely, regions of interest based feature extraction in facial emotion, sparse coding-SURF based feature extraction in body gesture and FCM based feature extraction in the speech emotion.

2.2 Facial Expression Feature Extraction

Before the feature extraction, we need to complete some prerequisite processes, such as the ROI region's segmentation of facial expression image, size adjustment, gray balance and so on.

In the face images, the texture and shape of the feature changes in three key parts-eyebrows, eyes and mouth may reflect changes in facial expression, as a consequence, we can use these ROI areas as facial image feature extraction's areas.

For facial images of JAFFE expression database one, manually obtains the coordinates of the four corners in three ROI areas, and split eyebrows, eyes and mouth from these primitive facial expression images. In Table 2.1, information about the coordinates of four corners in three ROI areas and rectangular clipping area are listed, each column in the four corners matrix coordinates represent four points x, y coordinates in clockwise direction; the rectangular clipping region matrix represent the width and height of this rectangular clipping area.

By clipping these ROI areas we can not only reduce the interference in facial information caused by image interference in noncritical part, but also reduce the amount of data and improve the computing speed. Specific ROI crop areas image is shown in Fig. 2.1.

Table 2.1 Comparison of experimental results of emotion recognition

Key parts	Corners coordinates	Clipping region
	$\begin{bmatrix} Xleftup & Xrightup & Xleftdown & Xrightdown \\ Yleftup & Yrightup & Yleftdown & Yrightdown \end{bmatrix}$	$[Length, Width]$
Eyebrows	$\begin{bmatrix} 74.21 & 182.26 & 182.26 & 74.21 \\ 100.14 & 100.24 & 120.05 & 120.05 \end{bmatrix}$	$\begin{bmatrix} 108.05 & 19.91 \end{bmatrix}$
Eyes	$\begin{bmatrix} 74.22 & 182.25 & 182.25 & 74.22 \\ 120.85 & 120.85 & 140.12 & 140.12 \end{bmatrix}$	$\begin{bmatrix} 108.03 & 19.27 \end{bmatrix}$
Mouth	$\begin{bmatrix} 95.55 & 162.01 & 162.01 & 95.55 \\ 180.03 & 180.03 & 210.05 & 210.05 \end{bmatrix}$	$\begin{bmatrix} 66.46 & 30.02 \end{bmatrix}$

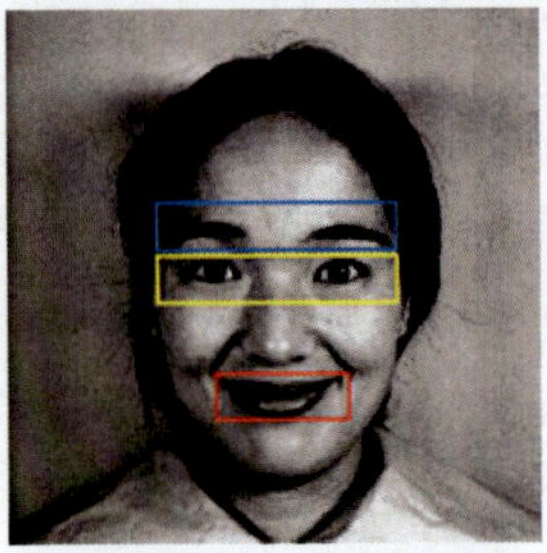

Fig. 2.1 Specific ROI crop areas

2.3 Speech Emotion Feature Extraction

With regard of speech emotional feature extraction, we adopt the non-personalization speech emotional feature based on derivative to supplement the traditional speech personalized emotional characteristics, and realize the universal and negotiability emotional characteristics. The speech emotional feature sets are computed using openSMILE toolkit (version 2.3). 16 basic features and their 1st derivatives are extracted as fundamental features. 16 basic features are F0, ZCR, RMS energy, and MFCC 1-12, respectively. Derivative features are less affected by different person, which are seen as non−personalized features, and 12 statistics values of these fundamental features are calculated. According to this method, the personalized features and non−personalized features are obtained.

In the speech emotional features, the zero-crossing rate (ZCR) refers to the fact that the signal pattern we are talking about passes through the zero-level record. The ZCR of the speech signal $\chi(m)$ is computed as

$$Z = \frac{1}{2}Z = \frac{1}{2}\sum_{m=0}^{N-1} |sgn[x(m)] - sgn[x(m-1)]| \tag{2.1}$$

MFCC can be calculated by the following steps:

Step 1: Improve the speaker's voice information by the Hamming window and the framing. After such pre-processing, the Fast Fourier Transformation (FFT) is implemented to obtain the spectrum.

Step 2: Square the result of *Step* 1, then pass it through the corresponding triangle filter, and then evenly arrange the center frequency according to the Mel frequency scale. The center frequency of the bandpass filter is divided by the interval of 150Mel and the bandwidth of 300Mel. Suppose the number of filters is M, and the output frequency after filtering is X(k), $k = 1, 2, \ldots, M$.

Step 3: Calculate the logarithm of the output of the bandpass filter in *Step* 2, and then calculate the obtained power log spectrum by the following formula to obtain K MFCC ($K = 12 - 16$), where K refers to the order of MFCC parameter. According to the symmetry, the transformation can be simplified as follows:

$$C_n = \sum_{k=1}^{K} \log Y(k) \cos[\pi(k - 0.5)n/N], n = 1, 2, \ldots, N \tag{2.2}$$

where N represents the number of filters, and C_n is the filtered output.

The personalized features have a beneficial effect on the SER of a certain person, but in the case of unfamiliar speaker without databases, the emotion recognition rate is not very high. Derivative-based non-personalized features can solve this problem. Therefore, both personalized features and non-personalized features are used for SER.

The Fuzzy c-means (FCM) is used for data clustering based on the method of bootstrap, the training sample data set D is obtained, i.e., $D = [y_1 \quad y_2 \quad \ldots \quad y_N^T]$ and $y_o = \left[y_{o1}\ y_{o2} \ldots y_{o384} \right], o = 1, 2, \ldots, N$, where N is the number of sample data. i represents the ith sample data. The FCM method is an iterative clustering algorithm to partition l normalized samples into c clusters by minimizing the following objective function

$$\begin{aligned} &\min J_m(U, V) = \sum_{k=1}^{L}\sum_{o=1}^{N} (\mu_{io})^m D_{ko}^2 \\ &D_{ko}^2 = \|y_o - c_k\|^2 \\ &\text{s.t.} \begin{cases} \sum_{k=1}^{L} \mu_{ko} = 1 \\ 0 < \mu_{ko} < 1 \end{cases} k = 1, \ldots, L, o = 1, \ldots, N \end{aligned} \tag{2.3}$$

where μ_{ko} is the membership value of the oth sample in the kth cluster, U is the related fuzzy partition matrix consisting of μ_{ko}, $V = (c_1, c_2, \ldots, c_L)$ is the cluster center matrix, L is the number of clusters, m is the fuzzification exponent which has an important regulatory effect on the fuzziness degree of clusters; we set usually $m = 2$. D_{ko} is the Euclidean distance between oth sample y_o and kth cluster center c_k.

2.4 Gesture Feature Extraction

In terms of gesture feature extraction, high-level gesture image representation is more significant for image understanding. Since the mapping of such low-level features as SURF and classification objects is a high nonlinear, it is difficult to directly construct the model for training, and achieve high classification performance. Therefore, it is necessary to build a salient high-level feature of the data. This section describes the feature extraction in body gesture in detail, which first perform foreground segmentation and SURF feature extraction, then sparse coding is used to extract deep features.

As far as RGB image segmentation is concerned, we adopt the iterative threshold segmentation method to extract the gesture foreground. The method could obtain a relatively accurate segmentation with less computational overhead using only a single parameter. The segmentation algorithm works well when the histogram of the image is bimodal, and the trough is deeper and wider. In the iterative threshold segmentation, Z_{max} and Z_{min} are the maximum and the minimum gray value of the image, the initial threshold T_0 is the average of Z_{max} and Z_{min}. The image is divided into the foreground and the background according to the threshold T_0, whose average gray value is Z_o and Z_b, respectively. So the new threshold is $T = \frac{(Z_o+Z_b)}{2}$. We repeat the above process until the following condition has been met:

$$|T_i - T_{i-1}| < \Delta T \tag{2.4}$$

where ΔT is a predefined threshold. The convergence condition is that the difference of the threshold after the iteration is less than a predefined threshold, which determines the accuracy of the threshold convergence.

In gesture feature extraction, we adopt the Speeded-up Robust Features (SURF) to detect the local features of images, to achieve the invariance for rotation, scale, lightness as well as the stability for perspective transformation, affine change and noise. The SURF algorithm can detect the features with high speed and a low computing overhead because of the using of integral images method. It uses a Hessian matrix to determine whether the points in the image are extreme points by searching for images in all scale spaces, which are potential points of interest that are not sensitive to scale and selection. For each point $f(x, y)$ of the image, the discriminant of the Hessian matrix is:

$$Det(H) = L_{xx} * L_{yy} - (L_{xy})^2 \tag{2.5}$$

where $L_{xx}(x, \delta) = \frac{\partial^2 g(\delta)}{\partial x^2} * I(x, y)$, which is the convolution of second-order Gaussian for integral image, and representing the Gaussian scale space of the image. δ denotes the scale parameter of the point (x, y).

In order to improve the calculation speed, the SURF algorithm uses a box filter instead of the Gaussian filter L:

$$Det(H) = D_{xx} D_{yy} - (w D_{xy})^2 \tag{2.6}$$

where $D_{(\cdot)}$ is the approximate solution corresponding to $L_{(\cdot)}$, and the value of w is set to 0.9. The role of this parameter is to balance the error caused by the use of the box filter. Different scale of images can be obtained by using different size box filter templates, which could help in search for the extreme point of the speckle response.

The algorithm uses a non-maximum suppression algorithm to determine the initial feature points, searching for the maximum value in the neighbourhood and surrounding scale space around each pixel, and filtering out weaker points and error-positioned points to screen for the final stable feature point. The main direction of the feature point is determined by the maximum value of the Haar wavelet feature in the circular neighborhood of the feature point.

For each point to be selected, we calculate the Haar wavelet features of 16 sub-regions in the square region around the feature points, and then get the following 4 values in each sub-region. The length of the eigenvector for each candidate point is $16 \times 4 = 64$.

$$\left[\sum dx, \sum dy, \sum |dx|, \sum |dy|\right] \tag{2.7}$$

The sparse coding (SC) algorithm is used to remove redundant features and acquire more significant gesture image features representation, which is benefit for the image classification and recognition. As an up-and-coming signal processing technology, sparse coding is an unsupervised learning algorithm which is used to find a set of super-complete basis vectors to represent the input vector as a linear combination of these base vectors.

Based on the method of SURF algorithm, the training sample data set X is obtained. Assuming the basis vectors are ϕ_i, and the over-complete dictionary is $\phi = \{\phi_i\}_{i=1}^{n} \in {}^{d \times n}$, then X can be represented as $X = \sum_{i=1}^{k} \alpha_i \phi_i$, in which α is the sparse coefficient. For an over-complete base, the coefficient α is typically under-determined by the input vector X. Therefore, the "sparsity" standard is introduced to solve the problem. The purpose of sparse coding algorithm is to minimize the following objective function:

$$a\hat{} = \min_{\alpha,\phi} \sum_{j=1}^{m} \left\| x^{(j)} - \sum_{i=1}^{k} \alpha_i^{(j)} \phi_i \right\|^2 + \lambda \sum_{i=1}^{k} S(\alpha_i^{(j)}) \tag{2.8}$$

$$subject \quad to \, \|\Phi\|^2 \leq C, \forall i = 1, \ldots, k \tag{2.9}$$

where λ is a regularization parameter which balances the influence of the reconstruction term and the sparsity term, m is the number of input vectors, k is the dimensionality of sparse coefficient, $S(.)$ is a sparse cost function. In practice, the general choice of the $S(.)$ is the $L1$ norm cost function $S(\alpha_i) = |\alpha_i|_1$ and the logarithmic cost function $S(\alpha_i) = \log(1 + \alpha_i^2)$.

SC has a training phase and a coding phase. Firstly, randomly initialize a visual dictionary ϕ and fix it, adjust α_i to minimize the objective function. Then the coefficient vector α is fixed, and we optimize the dictionary ϕ. Finally, iterating the former two steps until convergence, and obtaining the visual dictionary ϕ and the sparse code α of the training sample X. In the coding phase, for each image represented as a descriptor set X, the SC codes are obtained by optimizing Eq. 2.8 with respect to α only.

2.5 Summary

Multi-modal emotion feature extraction is an indispensable part of multi-modal emotion recognition. In this chapter, three methods are proposed for three modes' features, namely, regions of interest based feature extraction in facial expression, sparse coding-SURF based feature extraction in body gesture and FCM based feature extraction in the speech emotion. In order to make effective use of emotion information in multi-modal emotion recognition, the corresponding feature extraction method should be adopted according to the characteristics of emotion information of different modes.

References

1. C.A.R. Behaine, J. Scharcanski, Enhancing the performance of active shape models in face recognition applications. IEEE Trans. Instrum. Meas. **61**(8), 2330–2333 (2012)
2. Y. Cheon, D. Kim, Natural facial expression recognition using differential-AAM and manifold learning. Pattern Recognit. **42**(7), 1340–1350 (2009)
3. S. Taheri, Q. Qiu, R. Chellappa, Structure-preserving sparse decomposition for facial expression analysis. IEEE Trans. Image Process. **23**(8), 3590–603 (2014)
4. K. Wang, R. Li, L. Zhao, Real-time facial expressions recognition system for service robot based-on ASM and SVMs, in *Proceedings of 8th World Congress on Intelligent Control and Automation (WCICA)*, pp. 6637–6641 (2010)
5. Y. Tian, T. Kanade, J.F. Cohn, Recognizing action units for facial expression analysis. IEEE Trans. Pattern Anal. Mach. Intell. **23**(2), 97–115 (2001)
6. G. Littlewort, J. Whitehill, T. Wu, The computer expression recognition toolbox (CERT), in *Proceedings of 2011 IEEE International Conference on Automatic Face & Gesture Recognition and Workshops (FG 2011)*, pp. 298–305 (2011)
7. M.F. Valstar, M. Pantic, Fully automatic recognition of the temporal phases of facial actions. IEEE Trans. Syst. Man Cybern. Part B: Cybern. **42**(1), 28–43 (2012)
8. S. Berretti, A.D. Bimbo, P. Pala, A set of selected SIFT features for 3D facial expression recognition, in *Proceedings of 20th International Conference on Pattern Recognition (ICPR)*, pp. 4125–4128 (2010)
9. V. Le, H. Tang, T.S. Huang, Expression recognition from 3D dynamic faces using robust spatio-temporal shape features, in *Proceedings of 2011 IEEE International Conference on Automatic Face & Gesture Recognition and Workshops (FG 2011)*, pp. 414–421 (2011)
10. K. Mistry, L. Zhang, S.C. Neoh et al., A micro-GA embedded PSO feature selection approach to intelligent facial emotion recognition. IEEE Trans. Cybern. **47**(6), 1496–1509 (2017)
11. Z. Zhang, Feature-based facial expression recognition: sensitivity analysis and experiments with a multilayer perceptron. Int. J. Pattern Recognit. Artif. Intell. **13**(6), 893–911 (1999)
12. S. Liu, Y. Tian, Facial expression recognition method based on Gabor wavelet features and fractional power polynomial kernel PCA, in *Advances in Neural Networks*, pp. 144–151 (2010)
13. G. Zhao, M. Pietikainen, Dynamic texture recongnition using local binary patterns with an application to facial expression. IEEE Trans. Pattern Recognit. Mach. Intell. **29**(6), 915–928 (2007)
14. F. Xu, J. Zhang, J. Wang, Microexpression identification and categorization using a facial dynamics map. IEEE Trans. Affect. Comput. **8**(2), 254–267 (2017)
15. X. Pu, K. Fan, X. Chen et al., Facial expression recognition from image sequences using twofold random forest classifier. Neurocomputing **168**, 1173–1180 (2015)

16. Y. Yacoob, L. Davis, Recognizing human facial expressions from long image sequences using optical flow. IEEE Trans. Pattern Anal. Mach. Intell. **18**(6), 636–642 (1996)
17. K. Anderson, P.W. Owan, A real-time automated system for the recognition of human facial expressions. IEEE Trans. Syst. Man Cybern. Part B: Cybern. **36**(1), 96–105 (2006)
18. I. Kotsia, I. Pitas, Facial expression recognition in image sequences using geometric deformation features and support vector machines. IEEE Trans. Image Process. **16**, 172–187 (2007)
19. S. Park, D. Kim, Subtle facial expression recognition using motion magnification. Pattern Recognit. Lett. **30**(7), 708–716 (2009)
20. H. Soyel, H. Demirel, Optimal feature selection for 3D facial expression recognition using coarse-to-fine classification. Turk. J. Electr. Eng. Comput. Sci. **18**(6), 1031–1040 (2010)
21. M. Liu, S. Li, S. Shan, Deeply learning deformable facial action parts model for dynamic expression analysis, in *Proceedings of Asian Conference of Computer Vision*, pp. 143–157 (2014)
22. P. Liu, S. Han, Z. Meng et al., Facial expression recognition via a boosted deep belief network, in *Proceedings of IEEE Conference on Computer Vision and Pattern Recognition*, pp. 1805–1812 (2014)
23. B. Schuller, A. Batliner, S. Steidl et al., Recognising realistic emotions and affect in speech: state of the art and lessons learnt from the first challenge. Speech Commun. **53**(9), 1062–1087 (2011)
24. S.G. Koolagudi, K.S. Rao, Emotion recognition from speech using source, system, and prosodic features. Int. J. Speech Technol. **15**(2), 265–289 (2012)
25. H.G. Wallbott, Bodily expression of emotion. Eur. J. Soc. Psychol. **28**(6), 879–896 (1998)
26. M. De Meijer, The contribution of general features of body movement to the attribution of emotions. J. Nonverbal Behav. **13**(4), 247–268 (1989)
27. A.F. Bobick, Movement, activity and action: the role of knowledge in the perception of motion. Philos. Trans. R. Soc. Lond. **352**, 1257–1265 (1997)
28. A.F. Bobick, J.W. Davis, The recognition of human movement using temporal templates. IEEE Trans. Pattern Anal. Mach. Intell. **23**(3), 257–267 (2001)
29. S.J. Lin, M.H. Chao, C.Y. Lee et al., Human action recognition using motion history image based temporal segmentation. Int. J. Pattern Recognit. Artif. Intell. https://doi.org/10.1142/S021800141655017X
30. Q.X. Wu, Z.Y. Wang, F.Q. Deng et al., Discriminative two-level feature selection for realistic human action recognition. J. Vis. Commun. Image Represent. **24**(7), 1064–1074 (2013)
31. X.Y. Wu, X. Mao, L.J. Chen et al., Trajectory-based view-invariant hand gesture recognition by fusing shape and orientation. IET Comput. Vis. **9**(6), 797–805 (2015)

Chapter 3
Deep Sparse Autoencoder Network for Facial Emotion Recognition

Check for updates

Deep neural network (DNN) has been used as a learning model for modeling the hierarchical architecture of human brain. However, DNN suffers from problems of learning efficiency and computational complexity. To address these problems, deep sparse autoencoder network (DSAN) is used for learning facial features, which considers the sparsity of hidden units for learning high-level structures. Meanwhile, Softmax regression (SR) is used to classify expression feature. In this chapter, Softmax regression-based deep sparse autoencoder network (SRDSAN) is proposed to recognize facial emotion in human-robot interaction. It aims to handle large data in the output of deep learning by using SR, moreover, to overcome local extrema and gradient diffusion problems in the training process, the overall network weights are fine-tuned to reach the global optimum, which makes the entire depth of the neural network more robust, thereby enhancing the performance of facial emotion recognition.

3.1 Introduction

With the rapid development of the theory and technology in robotics, it is generally desired that robots could have the ability to recognize and understand human emotion. Meanwhile, the intelligent service system with emotional recognition ability becomes a hot topic in human-robot interaction. The facial expression plays an important role in manifestations of human emotional expression [1], in consequence the facial emotion recognition become the main subject in the field of affective computation.

During facial recognition, traditional facial feature extraction algorithms face main constraints just like face posture diversity and changeability, individual differences in facial structure and the levels of skin color, computer performance impose restrictions on training speed, the impact on the external environment, for instance, light, shelter and so on, such as optical flow method [2], model method [3, 4], track

L. Chen et al., *Emotion Recognition and Understanding for Emotional Human-Robot Interaction Systems*, Studies in Computational Intelligence 926,
https://doi.org/10.1007/978-3-030-61577-2_3

the point feature method [5, 6]. Therefore, artificial neural network technology [7] began to frantically sweep the worldwide in the field of human-robot interaction when the traditional way of feature extraction and recognition stepped into the bottleneck.

However, the traditional neural network algorithm is always prone to cause local maximum and gradient diffusion problems in the training process, in addition, training set need to rely on a large numbers of labeled samples, scarce samples will result in training over-fitting. In consequence, deep learning network [8] which is able to realize unsupervised self-learning without relying on a large number of labeled samples has become a hot research topic in current research. The "Google Brain" project built a machine learning model named deep neural networks (DNN), which has made great breakthroughs in image and speech recognition areas. According to the model, deep learning and training methods can be divided into convolution neural network (CNN) [9], the depth of belief networks (DBN) [10], restricted Boltzmann machine (RBM), etc.

Regression techniques, such as ridge regression (RR) [11, 12] and logistic regression (LR) [13, 14], have been widely used in supervised learning for pattern classification. In the recent year, ridge regression (RR) is generalized for face recognition [15–17]. In visual classification tasks such as face recognition, the appearance of the training sample images also conveys important discriminative information. In [15, 16], RR uses regular simplex vertices to represent the multiple target class labels, which generalize RR for multivariate labels in order to apply it for face recognition. Kernel ridge regression (KRR), which is the extension of RR, is used for multivariable nonlinear regression [18, 19]. The nonlinear structure of face images, is also proposed using the kernel trick [20, 21]. In [22], the combination of KRR and the truncated-regularized Newton method, which is based on the conjugate gradient (CG) method, leads to a powerful regression method. However, these methods mainly exploit the class label information for linear mapping function learning, and they will become less effective when the number of training samples per class is large. As a result, in this chapter, we apply Softmax regression for facial emotion recognition to handle the large images in deep learning. In order to achieve a powerful regression, we consider a joint approach [23–25]. Therefore, back propagation algorithm is used for fine-tuning the weights of the whole deep sparse autoencoder network to achieve the global optimum.

Softmax-regression-based deep sparse autoencoder network (SRDSAN) is proposed to recognize facial expression for overcoming the local extrema and gradient diffusion problems. Firstly, the regions of interest (ROI) include eyebrows, eyes, and mouth, are selected as facial image feature extraction's areas. Then, greedy pre-train network layer by layer to get initial weights [8], and expand the 'code' and 'decode' networks, furthermore, optimize the sparse parameter, the hidden layer nodes and the numbers of hidden layers to determine the best network model. The top-level network use Softmax regression to classify expression feature, and train this process by gradient descent method (GD) to find the optimal model parameters. Finally, the weights of the entire DSAN are fine-tuned via back-propagation (BP) algorithm, so as to make the whole deep learning network more robust and enhance facial emotion recognition performance. The algorithm use the layered approach to dispose training data and extract to different levels of data features, which build signal mapping feature from the bottom to the top. In addition, the most advantage of it that there is no need to rely on labeled sample data for training, and it can learn features automatically without supervision.

To verify the accuracy of emotion recognition, simulation experiments are developed in an emotion recognition system established by using MATLAB and VC. Moreover, preliminary application experiments are being carried out in the developing human-robot interaction system called emotional social robot system (ESRS), including two mobile robots with kinect. By using the ESRS, the proposal is being extended to mobile robots for analyzing and understanding human emotions, as well as responding appropriate social behaviors.

3.2 Softmax Regression Based Deep Sparse Autoencoder Network

Network structure of Softmax-regression-based deep sparse autoencoder network (SRDSAN) is shown in Fig. 3.1, which uses the sparse autoencoder network for deep learning, and Softmax regression to classify expression feature. The algorithm of SRDSAN is as follows.

Algorithm 3.1: The algorithm of SRDSAN.

1. The ROI areas clipping, and normalize the image $(x, y) \Rightarrow x$.
2. Greedy pre-train network layer by layer to get initial weights matrix w_i.
3. Obtain the network output $h_{w,b}(x)$.
4. Training Softmax regression to get the model parameters $\theta_1, \theta_2, \ldots, \theta_k$.
5. Minimize cost function $J(\theta)$.
6. Fine-tune the weights of the entire SRDSAN.
7. Obtain the facial emotion Y.

Algorithm 3.1 outlines the SRDSAN method. First, eyebrows, eyes, and mouth are selected as the ROI and extracted as feature of facial expression images. Then, initial weights of the network is produced by greedy pre-training the network layer by layer. To determine the best network model, hidden layer nodes and the numbers of hidden layers is obtained by optimizing the sparse parameter. Furthermore, SR is used to classify facial expression features, and the optimal model parameters of SR is trained by GD method. Finally, the weights of the entire DSAN are fine-tuned via BP algorithm, to make the whole deep learning network more robust and enhance facial emotion recognition performance.

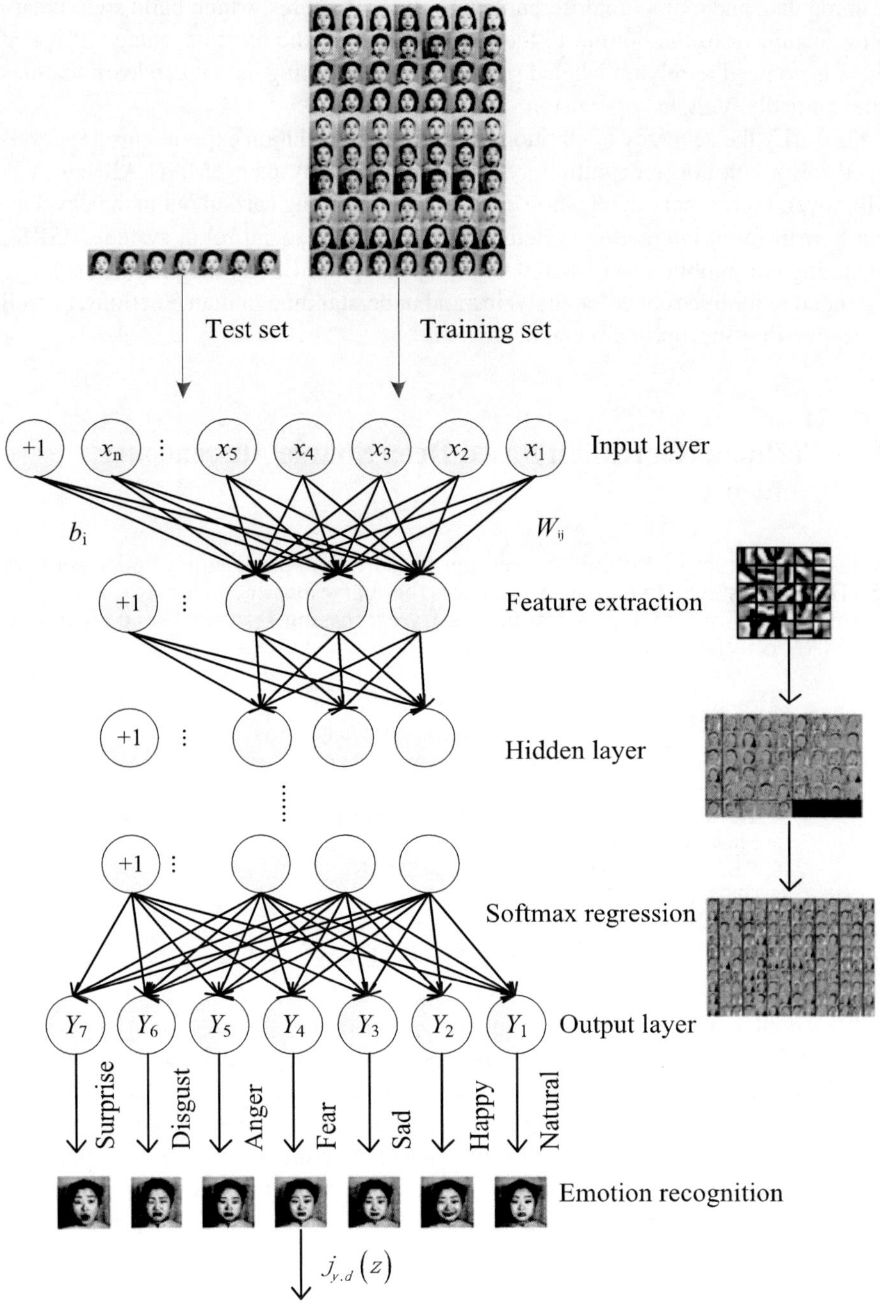

Fig. 3.1 Structure of SRDSAN

3.3 ROI Based Face Image Preprocessing

Before the feature extraction, we need to do some pretreatment processes, such as the ROI region's segmentation of facial expression image, size adjustment, gray balance and so on.

In the face images, the texture and shape of the feature changes in three key parts-eyebrows, eyes and mouth may reflect changes in facial expression, as a consequence, we can use these ROI areas as facial image feature extraction's areas.

For facial images of JAFFE expression database [26], we manually obtain the coordinates of the four corners in three ROI areas, and split eyebrows, eyes and mouth from these primitive facial expression images. In Table 3.1, information about the coordinates of four corners in three ROI areas and rectangular clipping area are listed, each column in the four corners matrix coordinates represent four points x, y coordinates in clockwise direction; the rectangular clipping region matrix represent the width and height of this rectangular clipping area.

By clipping these ROI areas can not only reduce the interference in facial information caused by image interference in noncritical part, but also reduce the amount of data and improve the computing speed. Specific ROI crop areas image are shown in Fig. 3.2.

Table 3.1 Comparison of emotion recognition experimental results

Key parts	Corners coordinates	Clipping region
	$\begin{bmatrix} Xleftup & Xrightup & Xleftdown & Xrightdown \\ Yleftup & Yrightup & Yleftdown & Yrightdown \end{bmatrix}$	$[Length, Width]$
Eyebrows	$\begin{bmatrix} 74.21 & 182.26 & 182.26 & 74.21 \\ 100.14 & 100.24 & 120.05 & 120.05 \end{bmatrix}$	$\begin{bmatrix} 108.05 & 19.91 \end{bmatrix}$
Eyes	$\begin{bmatrix} 74.22 & 182.25 & 182.25 & 74.22 \\ 120.85 & 120.85 & 140.12 & 140.12 \end{bmatrix}$	$\begin{bmatrix} 108.03 & 19.27 \end{bmatrix}$
Mouth	$\begin{bmatrix} 95.55 & 162.01 & 162.01 & 95.55 \\ 180.03 & 180.03 & 210.05 & 210.05 \end{bmatrix}$	$\begin{bmatrix} 66.46 & 30.02 \end{bmatrix}$

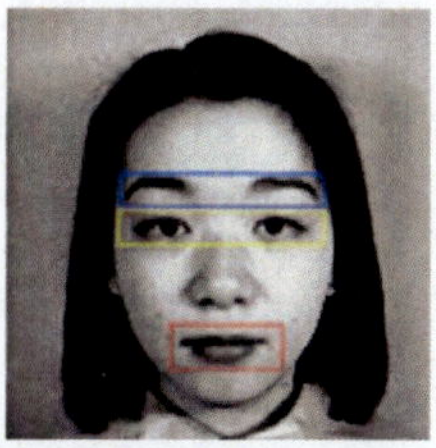

Fig. 3.2 The ROI segment region image

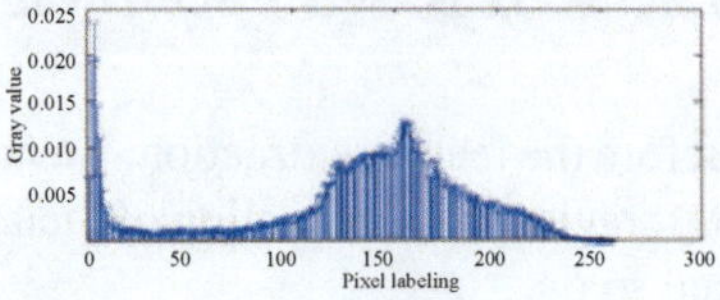

Fig. 3.3 Histogram before gray scale equalization

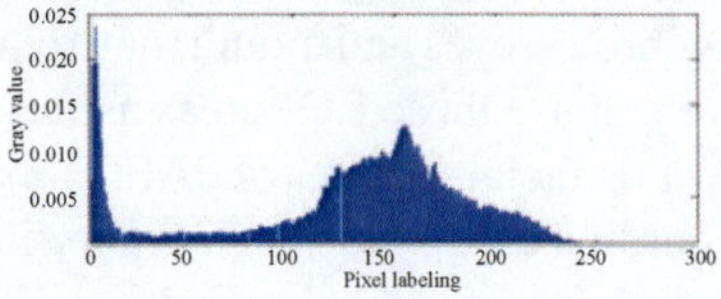

Fig. 3.4 The histogram after gray scale equalization

In the gray scale histogram of the original image, if the pixel values are concentrated in certain gray range, the image will not have a strong contrast. By adjusting the image's gray values, which will be evenly distributed throughout the range of the gray area, so as to make sure the gray zone have approximately the same number of pixels, thus guaranteeing the image contrast is enhanced. The gray levels histograms are shown in Figs. 3.3 and 3.4.

3.4 Expand the Encode and Decode Network

We define v as input layer data, define h as hidden layer data, use the trained parameters and input layer's data to calculate v and joint probability distribution function of h, and use this function's value as the initial matrix weight. Probability distribution function is described as follows [27],

$$p(h_j = 1|v) = \sigma\left(c_j + \sum_i w_{ij} v_i\right) \tag{3.1}$$

where the sigmoid function is as follows, and hide layer offset is

$$\sigma = \frac{1}{1 + \exp(x)} \tag{3.2}$$

The initial weighted matrix is defined as $w_i\,(i = 1, 2, \ldots n)$, and the network input data is defined as x, the network output data is defined as $h_{w,b}(x)$.

In the code phase, input data x through the mapping function is activated to give rise to u as follows,

$$u = g(w_i x + b_i) \tag{3.3}$$

and activation function $g(\cdot)$ is the sigmoid function,

$$g(\cdot) = \frac{1}{1 + e^{-x}} \tag{3.4}$$

In the decode stage, by activation function of reconstruct the input signal u, obtained as follows,

$$h_{w,b}(x) = g({w_i}^T u + b_{i+1}) \tag{3.5}$$

Softmax regression is used to sort the features learned by deep sparse autoencoder network, for training set: $\{(x^{(1)}, y^{(1)}), \ldots, (x^{(m)}, y^{(m)})\}$, it has $y^{(i)} \in \{1, 2, \ldots, k\}$.

Definition of $k = 7$ categories in facial emotion recognition, are natural, happiness, anger, sadness, surprise, disgust, fear, respectively. We define an assumed function to estimate the probability distribution for each category, the assumed function $h_\theta(x)$ using Eq. 3.5.

The model parameter matrix is defined as follows,

$$\theta = \begin{bmatrix} \theta_1^T \\ \theta_2^T \\ \vdots \\ \theta_k^T \end{bmatrix} \tag{3.6}$$

The Softmax regression's cost function adopts Eq. 3.5. We add decay value in the cost function in order to punish too large value of parameter, it can not only retain all the parameters like $\theta_1, \theta_2, \ldots, \theta_k$, but also solve the problem of redundant parameters. The cost function is rewritten as,

$$J(\theta) = -\frac{1}{m}\left[\sum_{i=1}^{m}\sum_{j=1}^{k} 1\{y^{(i)} = j\} \log \frac{e^{\theta_j^T x^{(i)}}}{\sum_{l=1}^{k} e^{\theta_j^T x^{(i)}}}\right] + \frac{\lambda}{2}\sum_{i=1}^{k}\sum_{j=0}^{n} \theta_{ij}^2 \tag{3.7}$$

We determine the derivative with respect to cost function, and update the parameters in each iteration,

$$\theta_j = \theta_j - \alpha \nabla_{\theta_j} J(\theta)(j = 1, 2 \ldots, k) \tag{3.8}$$

Using the above steps in an interative way we optimize the Softmax classification model to achieve an optimal regression model.

3.5 Softmax Regression

Here we present the Softmax classifier which is the expansion of the logical classifier. The logical classifier is more suitable for some nonlinear classification problems, and it is only suitable for the binary classification problem. The classification results are the categories of probability as output; the final category is determined by the threshold. The softmax classifier can expand the logic classifier, and make it have many classes of nonlinear classification ability.

The logical classifier calculates a probability value, and then compares with threshold ϕ, which can be converted into a simple binary classification problem. The expression of the logical function is shown in formula, where $h_\theta(x)$ is the probability of 1, and θ is the parameter of the model.

$$h_\theta(x) = g\left(\theta^T x\right)\frac{1}{1+e^{-\theta^T x}} = p\left(y = 1\,|x;\theta\right) \tag{3.9}$$

Training model parameters is by constantly adjusting and minimizing the loss function through training. The loss function is expressed in the form.

$$J(\theta) = -\frac{1}{m}\left[\sum_{i=1}^{m}\sum_{j=1}^{k} 1\{y^{(i)} = j\}\log\frac{e^{\theta_j^T x^{(i)}}}{\sum_{l=1}^{k} e^{\theta_j^T x^{(i)}}}\right] \tag{3.10}$$

The logical classifier can be extended to deal with many kinds of classifiers, this is the multiple softmax classifier. The expansion mode of the logical classifier is shown in the Eq. 11.3, where the output value is a k dimensional vector.

$$h_\theta(x^{(i)}) = \begin{bmatrix} p(y^{(i)} = 1|x^{(i)};\theta) \\ p(y^{(i)} = 2|x^{(i)};\theta) \\ \vdots \\ p(y^{(i)} = k|x^{(i)};\theta) \end{bmatrix} = \frac{1}{\sum_{j=1}^{k} e^{\theta_j^T x(i)}}\begin{bmatrix} e^{\theta_1^T x(i)} \\ e^{\theta_2^T x(i)} \\ \vdots \\ e^{\theta_k^T x(i)} \end{bmatrix} \tag{3.11}$$

where the model parameters are $\theta_1, \theta_2, \ldots, \theta_k$.

The classification result should only be a scalar, but the expanded output value is no longer a scalar, because of the expansion of the logical classifiers using the parallel model. For each category, j outputs a calculated probability value, which indicates the probability of the data object divided into this category. It has achieved the categorization with the category corresponding to the maximum probability value.

3.6 Overall Weight Training

Sample set's overall cost function that contains m samples can be expressed as follows,

$$J(w,b) = \left[\frac{1}{m}\sum_{i=1}^{m} J(w,b,x^{(i)},y^{(i)})\right] + \frac{\lambda}{2}\sum_{l=1}^{n_l-1}\sum_{i=1}^{s_l}\sum_{j=1}^{s_{l+1}}\left(w_{ji}{}^{(l)}\right)^2 \tag{3.12}$$

where the first item of the formula is a mean square error term, and the second item of the formula is a weighted decay.

Forward equation is employed to calculate of activation value in each layer of network (not including the output layer), such forward equation is,

$$a^{(l+1)} = f(w^{(l)}a^{(l)} + b^{(l)}) \tag{3.13}$$

Using recursive principle to calculate residual error one has,

$$\delta_i^{(n_l)} = \left(\sum_{j=1}^{s_{l+1}} w_{ji}^{l} \delta_i^{(l+1)} \right) f'(z_i^{n_l}) \tag{3.14}$$

For $i = 1, 2 \ldots, m$, at each iteration we have to update the weights, the perhinent computing steps are presented below

$$\begin{aligned} &\Delta w^{(l)} = \Delta w^{(l)} + a_j^{(l)} \delta_i^{(l+1)} \\ &w^{(l)} = w^{(l)} - a\left[\left(\tfrac{1}{m}\right) \Delta w^{(l)}\right] + \lambda w^{(l)} \\ &\Delta b^{(l)} = \Delta b^{(l)} + \delta_i^{(l+1)} \\ &b^{(l)} = b^{(l)} - a\left[\tfrac{1}{m} \Delta b^{(l)}\right] \end{aligned} \tag{3.15}$$

With the above iterative procedure, weight parameters are optimized, therefore, we determine the minimum cost function, thus arriving at the optimal network model.

3.7 Experiments

System workflow uses kinect that on the top of wheeled robot to track facial expression images firstly, then invokes facial emotion recognition algorithm to feature extraction and emotion recognition, which is designed by MATLAB, and relies on the affective computing workstations as shown in Fig. 3.5.

Fig. 3.5 Emotional computing workstation

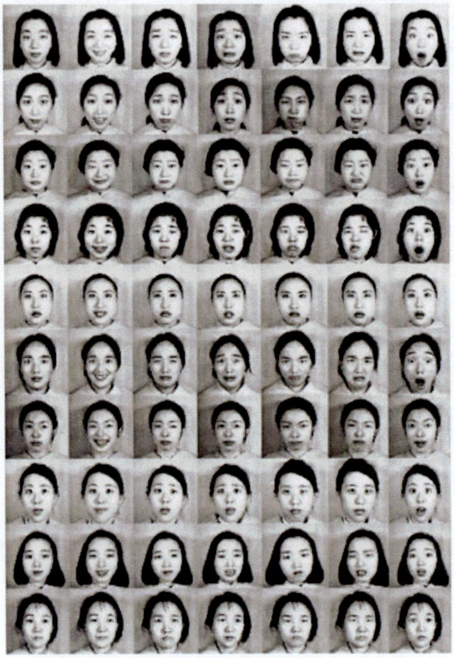

Fig. 3.6 Samples of the JAFFE facial expression library

Moreover, JAFFE facial expression database from Japan is used as training sample, which contains 213 facial expression images, including ten subjects, seven types of basic expressions. The sample images of JAFFE are shown in Fig. 3.6.

3.7.1 Fine-Tune Effect on Performance of Recognition

Visualization of underlying characteristics for weights that learned by sparse autoencoder network is designed, and the numbers of neurons node in the hidden layer are set to 140 to obtain an initial image of characteristics' visualization. Visual characteristics weight matrix is shown as follows.

Figures 3.7 and 3.8 present characteristic matrix of weights visualization images before and after fine-tuning overall weights, it can be seen that the features self-learned by overall network looks more sophisticated after fine-tuning the weights to ensure high recognition accuracy. The relation curve between the seven kinds of facial emotion recognition rate and the sparsity parameter is shown in Fig. 3.9.

According to the comparison of the Fig. 3.10 with Fig. 3.11, we conclude that fine-tuning can make the overall cost function converge faster, and in the actual test of 182 times when they have converged and stopped training.

Fig. 3.7 The weights visualization of the underlying characteristics

Fig. 3.8 The fine-tuning the weights visualization of the underlying characteristics

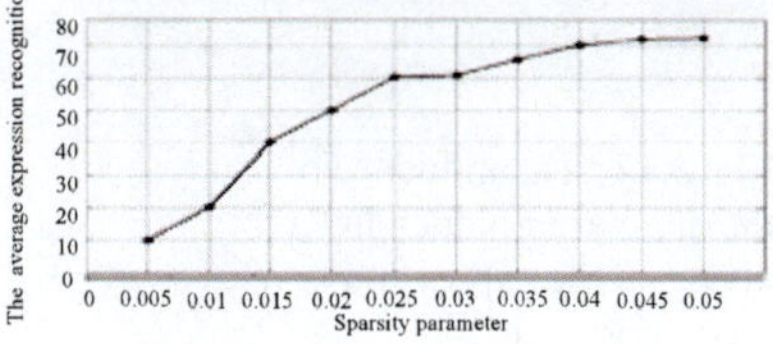

Fig. 3.9 The influence of sparse parameter and fine-tuning the weights on the rate of facial emotion recognition

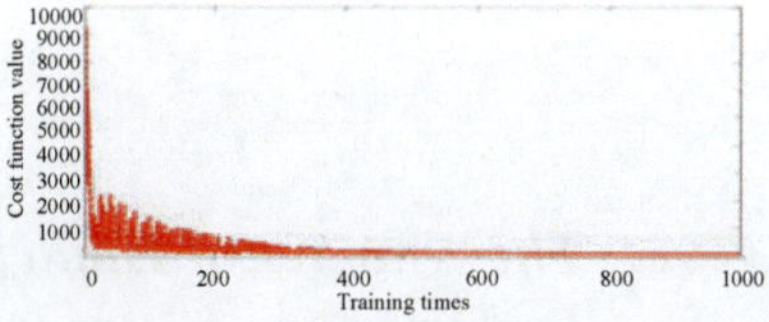

Fig. 3.10 The convergence of overall cost function before fine-tuning

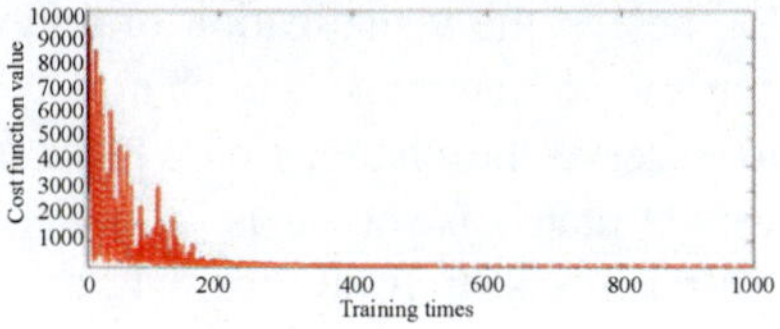

Fig. 3.11 The convergence of overall cost function after fine-tuning

Fig. 3.12 Weights visualizations for the hidden layer node number is 50

Fig. 3.13 Weights visualizations for the hidden layer node number is 200

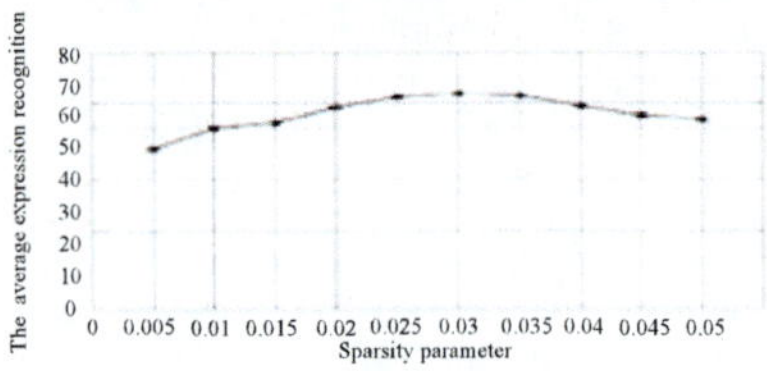

Fig. 3.14 The influence of the number of hidden layer nodes on the expression recognition rate

3.7.2 The Number of Hidden Layer Node's Effect on Performance of Recognition

We design the visualization of underlying characteristic for weights in the sparse autoencoder network, and change the numbers of neurons node in the hidden layer to observe the changes of weight matrix's feature images. Visual characteristics weight matrix is shown as follows.

It can be seen from Figs. 3.12 and 3.13 that the increase of the numbers of hidden layer nodes can increase the recognition rate of emotion, however, too large numbers of hidden layer nodes is helpless to improve the recognition rate of emotion. It will cause over fitting of the network and is not conducive to expression feature extraction. After fine-tuning the weight matrix, the recognition rate will be improved to a certain extent, and offset the impact of the hidden layer nodes numbers change, shown in Fig. 3.14.

Table 3.2 Comparison of emotion recognition experimental results

Emotion	Softmax regression (%)	SRDSAN (%)
Natural	86.667	100.00
Happy	80.000	93.333
Sad	63.333	100.00
Fear	63.333	76.667
Angry	83.333	100.00
Disgust	66.667	93.333
Surprise	70.000	100.00
Average	73.333	94.761

3.7.3 Recognition Rate

To verify the accuracy of emotion recognition, simulation experiments are completed by using MATLAB, with the results listed in Table 3.2.

According to the table, the average recognition accuracy of Softmax regression is 73.33% in the final test. Nevertheless, if we use unlabeled training data to train deep autoencoder network firstly, and then train the Softmax regression model, and we can find that the number of iteration convergence are only 181, and the accuracy is 94.76%. By fusing the Softmax retrogression in deep learning, it can be seen that the features self-learned by overall network looks more sophisticated after fine-tune, and fine-tune makes the overall cost function converge faster, which overcomes the local extrema and gradient diffusion problems. Moreover, this shows that the characteristics learned by our self-learning sparse autoencoder network are more representative than the characteristics of the original input data, which is the typical difference between conventional training methods and deep learning training methods. In addition, texture and shape feature changes in the three key parts (region of interest, ROI) such as eyebrows, eyes, mouth may reflect exactly the changes in facial expression features.

3.8 Summary

Softmax-regression-based deep sparse autoencoder network (SRDSAN) is proposed for facial emotion recognition, where sparse autoencoder and Softmax regression are employed to deep learning and recognize facial expression.

To verify the accuracy of the proposal, preliminary experiments completed in MATLAB by using standard Japanese facial expression library named JAFFE. The average emotion recognition reached 73.333% using Softmax regression model, but if we use the proposed SRDSAN, the average emotion recognition rate elevates to

94.761%. According to the experimental results, by fusing the Softmax retrogression in deep learning, it can be seen that the features self-learned by overall network looks more sophisticated after fine-tune, and fine-tuning makes the overall cost function converge faster, which overcomes the local extrema and gradient diffusion problems.

References

1. L. Zhang, D. Tjondronegoro, Facial expression recognition using facial movement features. IEEE Trans. Affect. Comput. **2**(4), 219–229 (2011)
2. C.K. Hsieh, S.H. Lai, Y.C. Chen, An optical flow-based approach to robust face recognition under expression variations. IEEE Trans. Image Process. **19**(1), 233–240 (2010)
3. Z. Li, D. Gong, X. Li, D. Tao, Aging face recognition: a hierarchical learning model based on local patterns selection. IEEE Trans. Image Process. **25**(5), 2146–2154 (2016)
4. M.A.A. Dewana, E. Grangerb, G.-L. Marcialisc, R. Sabourinb, F. Rolic, Adaptive appearance model tracking for still-to-video face recognition. Pattern Recognit. **49**, 129–151 (2016)
5. E. Vezzetti, F. Marcolin, G. Fracastoro, 3D face recognition: an automatic strategy based on geometrical descriptors and landmarks. Robot. Auton. Syst. **62**(12), 1768–1776 (2014)
6. L. Xu, W.W. Liu, K. Tsujino, C.W. Lu, A facial recognition method based on 3-D images analysis for intuitive human-system interaction, in *Proceedings of International Joint Conference on Awareness Science and Technology and Ubi-Media Computing (iCAST-UMEDIA)*, Aizuwakamatsu, Japan, pp. 371–377 (2013)
7. B.J.T. Fernandes, G.D.C. Cavalcanti, T.I. Ren, Face recognition with an improved interval type-2 fuzzy logic sugeno integral and modular neural networks. IEEE Trans. Syst. Man Cybern.-Part A: Syst. Hum. **41**(5), 1001–1012 (2011)
8. G.E. Hinton, R.R. Salakhutdinov, Reducing the dimensionality of data with neural networks. Science **313**(5786), 504–507 (2006)
9. Y. Sun, Y. Chen, X. Wang et al., Deep learning face representation by joint identification-verification, in *Advances in Neural Information Processing Systems*, vol. 27, pp. 1988–1996 (2014)
10. P. Liu, S. Han, Z. Meng et al., Facial expression recognition via a boosted deep belief network, in *Proceedings of IEEE Conference on Computer Vision and Pattern Recognition*, pp. 1805–1812 (2014)
11. P.Y. Wu, C.C. Fang, J.M. Chang, S.Y. Kung, Cost-effective kernel ridge regression implementation for keystroke-based active authentication system. IEEE Trans. Cybern. PP(99), 1–12 (2016)
12. J. García, R. Salmerón, C. García et al., Standardization of variables and collinearity diagnostic in ridge regression. Int. Stat. Rev. **84**(2), 245–266 (2015)
13. O. Ouyed, M.S. Allili, Feature relevance for kernel logistic regression and application to action classification, in *Proceedings of IEEE International Conference on Pattern Recognition*, pp. 1325–1329 (2014)
14. N. Herndon, D. Caragea, A study of domain adaptation classifiers derived from logistic regression for the task of splice site prediction. IEEE Trans. NanoBioscience **15**(2), 75–83 (2016)
15. S. An, W. Liu, S. Venkatesh, Face recognition using kernel ridge regression, in *Proceedings of IEEE Conference on Computer Vision and Pattern Recognition*, pp. 1–7 (2007)
16. H. Xue, Y. Zhu, S. Chen, Local ridge regression for face recognition. Neurocomputing **72**(4), 1342–1346 (2009)
17. G. Gao, J. Yang, S. Wu et al., Bayesian sample steered discriminative regression for biometric image classification. Appl. Soft Comput. **37**(C), 48–59 (2015)
18. P. Exterkate, P.J.F. Groenen, J.C. Hei et al., Nonlinear forecasting with many predictors using kernel ridge regression. Creat. Res. Pap. **32**(3), 736–753 (2013)

19. J. Li, L. Su, C. Cheng, Finding pre-images via evolution strategies. Appl. Soft Comput. **11**(6), 4183–4194 (2011)
20. C. Saunders, A. Gammerman, V. Vovk, Ridge regression learning algorithm in dual variables, in *Proceedings of the 15th International Conference on Machine Learning (ICML98)*, Madison-Wisconsin, pp. 515–521 (1998)
21. J. Li, L. Su, C. Cheng, Finding pre-images via evolution strategies. Appl. Soft Comput. **11**(6), 4183–4194 (2011)
22. M. Maalouf, D. Homouz, Kernel ridge regression using truncated newton method. Knowl.-Based Syst. **71**, 339–344 (2014)
23. M. Gong, J. Liu, H. Li et al., A multiobjective sparse feature learning model for deep neural networks. IEEE Trans. Neural Netw. Learn. Syst. **26**(12), 3263–3277 (2015)
24. W. Huang, L. Xiao, Z. Wei et al., A new pan-sharpening method with deep neural networks. IEEE Geosci. Remote Sens. Lett. **12**(5), 1037–1041 (2015)
25. Z. Zhang, J. Li, R. Zhu, Deep neural network for face recognition based on sparse autoencoder, in *Proceedings of International Congress on Image and Signal Processing* (2015)
26. M.J. Lyons, J. Budynek, S. Akamatsu, Automatic classification of single facial images. IEEE Trans. Pattern Anal. Mach. Intell. **21**(12), 1357–1362 (1999)
27. G.E. Hinton, Distributed representations. Technical Report (University of Toronto, 1984)

Chapter 4
AdaBoost-KNN with Direct Optimization for Dynamic Emotion Recognition

AdaBoost-KNN using adaptive feature selection with direct optimization is proposed for dynamic emotion recognition in human-robot interaction, where the real-time dynamic emotion is recognized based on facial expression. It can make robots capable of understanding human dynamic emotions, in such a way that human-robot interaction is realized in a smooth manner. Based on the facial key points extracted by Candide-3 model, adaptive feature selection is adopted, namely Plus-L Minus-R Selection is completed. It can determine the features that contribute the most to emotion recognition, thereby forming the basis of emotion classification. Emotion classification is based on AdaBoost-KNN, which builds a series of KNN classifiers. AdaBoost-KNN adjusts the weights of the data in an iterative manner. Moreover, global optimal parameters are approximated with direct optimization until the recognition rate reaches its maximal value.

4.1 Introduction

With the development of various technologies, the degree of social intelligence is increasing at the same time, and people naturally expect the robot to exhibit this emotional ability [1]. However, existing machines have not yet been able to communicate with people in a smooth and emotional way [2]. Studies show that up to 55% of human emotion is expressed by facial emotion [3, 4]. Facial emotion recognition exhibiting the vital significance in human-robot interaction [5, 6], has also been developing rapidly in recent years. Thus, the realization of facial emotion recognition will help the robot to understand human emotion and even emotional intention.

Most of the research has been focused on static image recognition, using geometric features, wavelet transform method based on Gabor, local binary model algorithm, and method based on hybrid feature extraction [7]. In [8], principal component analysis combined with logistic regression analysis was presented to deal

L. Chen et al., *Emotion Recognition and Understanding for Emotional Human-Robot Interaction Systems*, Studies in Computational Intelligence 926,
https://doi.org/10.1007/978-3-030-61577-2_4

with image information. Some disadvantages of these methods are that they lack sufficient dynamic information, and they are easily affected by the environment and individual differences [9]. There are some studies on dynamic facial emotion [10], commonly including optical flow method, feature point tracking method, motion model algorithm, and others. Reference [11] can be taken into account, where the facial emotion recognition in image sequences is realized by using support vector machines (SVMs), the [12] is based on geometric deformation features. Most of the existing algorithms focus on the sequence of dynamic images, in which real-time dynamic emotion recognition cannot be directly realized. Under this consideration, we use Candide3-based feature point matching method, which cannot only obtain the feature point, but also the dynamic information.

On the basis of feature extraction, researchers have completed a lot of work in feature selection [13]. Feature selection can be divided into Embedded, Filter and Wrapper [14]. A Particle Swarm Optimization (PSO) variant called chaotic mixing-based PSO is proposed to identify the most discriminative bodily characteristics [15]. Alweshah and Abdullah [16] proposed two hybrid Flux Approximation (FA) to optimize the weights of a probabilistic neural network (NN). The [17] proposed a modified FA incorporating opposition-based and dimensional-based methodologies (ODFA), to deal with high dimensional optimization problems. However, filtering and embedded feature selection usually comes with high computational complexity, long execution time and poor generalization. Therefore, we chose the adaptive feature selection (AFS) algorithm based on the Plus-L Minus-R Selection (LRS) method, which can adaptively adjust the parameters of the feature selection algorithm according to the recognition results obtained each time.

Facial emotion recognition methods make use of traditional and non-traditional classifiers [18]. The traditional classifiers include artificial neural network (ANN), hidden Markov model (HMM) [19], SVM [20], and K-Nearest Neighbour (KNN), etc. Many completed analysis of the characteristics and distribution of facial empression features [21], and Improved KNN classifier is used to enhance the performance of the classifier. Sebe obtained high accuracy using KNN algorithm on based on geometric features [22]. KNN classification as dynamic feature classifier in [23] was used. In [24] Curvature-KNN algorithm to realize 3D face recognition was adapted. SVM and Deep method have a long training and testing time. SVM and Deep method have a long training and testing time. In addition, in the multi-classification problem, the large amount of secondary programming calculation of SVM leads to large amount of classification calculation and slow classification speed. As for deep method, taking CNN as an example, it entails a large amount of data and a large number of iterations for training, which requires a high level of computing overhead. The training time of deep learning method is also long. KNN has the advantage that it can be calculated with time series and its training cycle is short. However, KNN still cannot solve the problem that some samples are weak or not representative of the given category. In order to make up for KNN's inability to adjust the weight of samples, we consider applying AdaBoost-KNN to recognize facial emotion states, such as angry, fear, happy, neutral, sad, and surprise [25]. In addition, the direct optimization was adopted to find the parameters with the highest recognition rate in the algorithm, which can automatically adjust the value of k and the number of iterations of the classifier until the optimal parameters were obtained.

The main idea of adaptive feature selection-based AdaBoost-KNN with direct optimization (AFS-AdaBoost-KNN-DO) is that the method produces emotional features based on the dynamic information of facial emotion. AFS is used for feature selection. AdaBoost-KNN completes the online dynamic facial emotion recognition, and finally produces results of facial emotion recognition. Specifically, at first, using Candide-3 based dynamic feature point extraction to get the coordinates of facial emotion point information and generate characteristic matrix. Considering that some characteristics of the contribution are not high, AFS is adopted to select the feature matrix by column. AdaBoost-KNN adjusts the weights of the training data. Then, the feature matrix is adjusted row by row with the iteration of AdaBoost-KNN. In such a way that the training data of emotion classification is adjusted row by row and column by column. Therefore, the feature selection and the successive classification is a complete unit and the results are obtained. Finally, the parameters in AdaBoost-KNN are adjusted according to the recognition rate using the direct optimization method, so that the recognition rate reaches the highest value.

The effectiveness of the proposal is validated through preliminary application experiments, and dynamic emotion recognition is established in the emotional social robot system. In the application experiment, two mobile robots were used as emotional presenters, and 10 graduate students were invited as volunteers.

4.2 Dynamic Feature Extraction Using Candide-3 Model

The Candide-3 model is the third generation of Candide model, which is a parameterized model [26]. It is made up of 113 points P_i ($i = 1, 2, \ldots, 113$). These dots can be linked into triangular meshes, each of which is referred to as a slice, with a total of 184 pieces. Candide-3 model gives 12 shape units and 11 action units controlling the change of the model. Candide-3 model gives the corresponding AUs unit and the corresponding expression feature point coordinate according to various sports unit, so it provides a convenient framework for facial expression analysis work. The model can be described as [27]:

$$g = sR(\bar{g} + AT_\alpha + ST_s) + t \tag{4.1}$$

where s is the amplification coefficient, $R = R(r_x, r_y, r_z)$ is the rotation matrix, g is the standard model, A is the moving unit, S is the shape unit, T_α and T_s are the corresponding change parameters. $t = t(t_x, t_y)$ is the transformation vector of the model in space, g is the expected face model. In the tracking process, the other parameters constitute vector b except the shape parameters. Therefore, the feature point tracking extraction algorithm based on Candide-3 model is essentially completed by updating b.

For an initial parameter b, we calculate the corresponding residual $r(b)$, and error $e(b)$:

$$r(b) = x' - x \tag{4.2}$$

$$e(b) = ||r(b)|| \tag{4.3}$$

where parameter Δb is obtained by multiplying residual image r by the update matrix:

$$\Delta b = -G'r = -(G^T G)^{-1} G^T r \tag{4.4}$$

where G is r-related gradient matrix, while new model parameters and model error can be obtained by calculating delta b:

$$b' = b + \rho \times \Delta b \tag{4.5}$$

$$e = e(b) \tag{4.6}$$

where ρ is a positive real number, and if $e' < e$, that parameter b is update according to the above formula until stable. If $e' \leq e$, reduce the ρ. When the error does not change, it is considered to be stable.

Our experiment relies on Kinect to verify the effect of facial feature point tracking in Candide-3 model. The experiment shows that even in the poorly illuminated environment, the tester presenting different head posture, Candide-3 model can well realize the face emotional feature point tracking.

4.3 Adaptive Feature Selection Based on Plus-L Minus-R Selection

In the practical application of machine learning, the number of features is usually large, among which there may be irrelevant features [28]. Feature selection can eliminate irrelevant or redundant features [29], so as to reduce the number of features, improve model accuracy, and reduce running time. Combined with sequence forward selection and sequence backward selection, LRS algorithm of heuristic search can solve the problems.

Once feature extraction has been completed, the feature selection algorithm starts from the empty set, adding L features in each round, and then eliminating R features one by one to achieve the optimal value of the evaluation function. Where L and R represent the number of features selected and eliminated at each round, and D represents the number of features left by the final cycle. In this chapter, the LRS algorithm starts from empty set and selects L groups of features from 113 sets of feature points to add them to feature set X successively. The distance of least squared error of the newly added feature groups and other features are calculated successively. The features of R groups that contribute little to the recognition results are removed from the selected feature set. After each cycle, the remaining L-R groups features that contribute most to emotional recognition were selected. In the actual test, the value of L is one to the maximum number of feature points, and the value of R is one to the maximum number of feature points minus one. Finally, feature selection is carried out through the cycle. When the system recognition rate reaches the highest value, we stop the selection procedure and obtain the required D groups feature data, forming a feature subset. The distance of least squared error of the newly added feature groups and other features is calculated as follows:

$$d = d_0 - \sum (x - x_i)^2 \tag{4.7}$$

where d is the distance, d_0 is the initial distance, i runs from 1 to N, and N is the number of feature points involved in the calculation, the value of newly sample feature points is x, and the value of other samples is x_i.

4.4 AdaBoost-KNN Based Emotion Recognition

AdaBoost-KNN is an integrated classifier, consists of a set of KNN (test data, train data, train label, k), where test data and train data are independent eigenvectors. Train label is the label for training sample data [30]. k = 1, 2,..., N represents the k samples that are most similar to the test samples located in the feature space. In the given facial feature data, the result of each KNN classification determines the optimal classification result. The steps to establish the classifier are as follows:

Step 1: Initialize weights of training data;

Step 2: Using the training data set with weight coefficient $D(m)$ to create the weak classifier $G_m(x)$, calculate the classification error rate of $G_m(x)$ on the training data set, and calculate the coefficient of $G_m(x)$;

Step 3: Update the data centralization value distribution, and built the final classifier according to the coefficient combination basic classifier;

Step 4: Train classifier using 70% data of JAFFE for training, and 30% data is used for testing. As a result, a total of 7 classifiers are trained;

Step 5: Use AdaBoost-KNN for classification of seven basic emotion. The decision-making process is as follows:

$$D_1 = (\omega_{11}, \ldots, \omega_{1i}, \ldots, \omega_{1N}) \tag{4.8}$$

$$\omega_{1i} = \frac{1}{N}, i = 1, 2, \ldots, N \tag{4.9}$$

where α_{1N} is the initial weight of the training sample. Each training sample is given the same weight at the beginning: $1/N$, and N is the number of training samples, and $D(m)$ is used to store the array of ω_{mi}. m = 1, 2,..., M represents the first number of iterations and the number of weak classifiers. Using the training data set with weight coefficient $D(m)$ to create the weak classifier $G_m(x)$, calculate the classification error rate e_m of $G_m(x)$ on the training data set, and calculate the coefficient ω_m of $G_m(x)$;

$$e_m = \sum_{i=1}^{N} \omega_{mi} I(G_m(x_i) \neq y_i) \tag{4.10}$$

$$\alpha_m = \frac{1}{2} log \frac{1 - e_m}{e_m} \tag{4.11}$$

Then update the data centralization value distribution.

$$\omega_{m+1,i} = \frac{\omega_{mi}}{Z_m} exp\left(-\alpha_m y_i G_m(x_i)\right) i = 1, 2, \ldots, N \tag{4.12}$$

$$D_{m+1} = \left(\omega_{m+1,1}, \omega_{m+1,2}, \ldots, \omega_{m+1,N}\right) \tag{4.13}$$

$$Z_m = \Sigma_{i=1}^{N} \omega_{mi} exp\left(-\alpha_m y_i G_m(x_i)\right) \tag{4.14}$$

The significance of Z_m is to make the sum of the weighting factor to be equal to 1.0, so that vector D is a probability distribution vector. Then we combine each weak classifier to get the final classifier. α_m is increased with the decrease of e_m, which means that the smaller the classification error rate is, the greater the role of the classifier in the final classifier becomes

$$G(x) = sign(f(x)) = sign\left(\Sigma_{m=1}^{M}(\alpha_m G(x))\right) \tag{4.15}$$

Each weak classifier is a separate KNN classifier [28]. The implementation steps of KNN are as follows:

Step 1: Calculate the Euclidean distance between the points in the known category data set and the test sample points;

Step 2: Sort the training sample points according to the increasing order of distance;

Step 3: Select the k training sample points with the minimum distance from the current point;

Step 4: Determine the occurrence frequency of the category of k points, and return the category with the highest frequency of k points as the prediction classification of the test point.

The Euclidean metric (Euclidean distance) is a commonly used distance [31], refers to the real distance between two points, or natural vector length in the dimensional space. In the N dimensional Euclidean space, vector x can be expressed as $x = (x_1, x_2, \ldots, x_n)$ is the real number, which is called the ith coordinate of x. The distance $\rho(A, B)$ between two points $A = (a_1, a_2, \ldots, a_n)$ and $B = (b_1, b_2, \ldots, b_n)$ is defined as follows:

$$\rho(A, B) = \sqrt{\Sigma\,(a_i - b_i)^2} \quad (i = 1, 2, \ldots, n) \tag{4.16}$$

The natural length of vector $x = (x_1, x_2, \ldots, x_n)$ is defined as:

$$|x| = \sqrt{x_1^2 + x_2^2 + \cdots + x_n^2} \tag{4.17}$$

The number of nearest neighbors is k. k can reflect the complexity of the whole model. The smaller the k is, the higher the complexity of the model becomes, and it is easy to be over-fitting. In application, the greater the difference between sample data and the smaller k is usually used. Generally, the optimal k is selected by cross validation.

4.5 AdaBoost-KNN with Direct Optimization for Emotion Recognition

The direct optimization (DO) method finds a group of optimal variables in the variable space through direct search, and it is one of the numerical methods. In essence, during the whole training process, the single weak classifier will exhibit some classification error, no matter how many rounds of training it underwent. However, the whole AdaBoost framework is likely to converge rapidly. Therefore, it is particularly important to select the basis for determining the end of iteration for the search of optimal parameters.

After each round of training, the AdaBoost adjusts the weight of samples, and the result of this adjustment is that the more samples are misclassified later, the higher the weight of samples will be. In this way, a single weak classifier will achieve a low weighted classification error by classifying the samples with high sample weight correctly. Although a single weak classifier can still cause classification errors on its own, the samples that are misclassified are all of low weight, and the final output of the AdaBoost framework will be "balanced" by the high weight weak classifier correctly classified earlier.

Record the number of iterations and emotion recognition error rate e(i) after each round of iteration, then, complete the next iteration when the algorithm adjust the weight of samples and classifier constantly. When iterative process has been completed, record the number of iterations $i + 1$ and affective recognition error rate e(i). If e(i) > e($i + 1$), records i+1 and e(i+1). Through successive iterations, Until the error rate of AdaBoost-KNN algorithm no longer decreases, the convergence condition is reached. Finally, Out of the loop, the parameters i and e($i + 1$) obtained are the optimal parameters applicable to the current model.

Therefore, when designing the convergence condition of emotion recognition, we adopted the method of recording the peak recognition rate, and The steps are as follows:

Step 1: Calculate the facial emotion recognition rate $1 - e(i)$;

Step 2: Increase the value of k and start the next round of calculation;

Step 3: Calculate the facial emotion recognition rate $1 - e(i + 1)$;

Step 4: Record the recognition rate obtained by the two rounds of calculation. If the recognition rate starts to decrease, stop the loop, and record the value of k and the recognition rate at this time, that is, the optimal two parameter values; otherwise, increase k and continue the algorithm.

The steps to determine the number of experimental iterations are as follows:

Step 1: Calculate the error rate of facial emotion recognition;

Step 2: Increase the i and start the next round of calculation;

Step 3: Calculate the error rate of facial emotion recognition $e(i + 1)$;

Step 4: Record the error rate of facial emotion recognition obtained by the two rounds of calculation. If the recognition rate starts to increase, stop the loop, and record the value of i and the recognition rate at this time, that is, the optimal two parameter values; otherwise, increase i and continue the algorithm.

4.6 Experiments

JAFFE database was selected to verify the effectiveness of the proposed algorithm. Meanwhile, control variable experiment was designed to explore the influence of the value of each parameter in the AFS-AdaBoost-KNN-DO algorithm on the recognition effect. Finally, the dynamic face emotion recognition algorithm is completed on the emotional social robot interaction system, and the results are analyzed on the basis of the application.

4.6.1 Experimental Environment and Data Selection

We set up an emotional social robot interaction system for the recognition algorithm, as shown in Fig. 4.1.

It is composed of a mobile robot, emotional computing workstation, router and data transmission equipment. The emotional workstation is configured to use Intel i5-4590 CPU with a CPU frequency of 3.3 GHz and a memory (RAM) of 4.00 GB and a system type of 64 bit operating system. The experimental software is MATLAB R2015b, and the corresponding simulation experiments are designed to verify the effectiveness of the algorithm. The database of facial emotion used in this experiment is the JAFFE public database [32].

The JAFFE database is composed 213 grayscale images made up of seven basic expressions of 10 women. The image size is 2 to 4 images per person. 70% of the images were selected as the training set, the other 30% of data forms a testing set. The example sample images of the database are shown in Fig. 4.2.

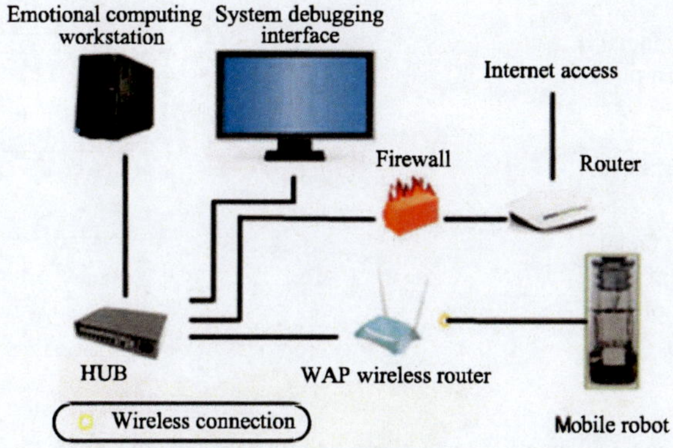

Fig. 4.1 The framework of emotional social robot system

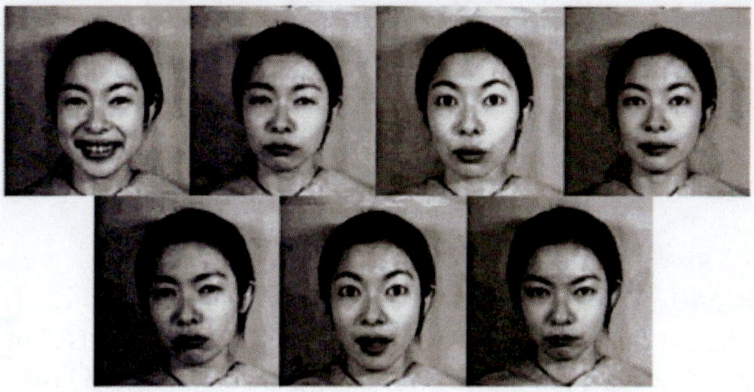

Fig. 4.2 Partial sample image of the database

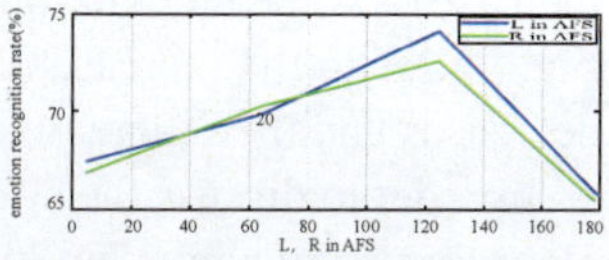

Fig. 4.3 Dynamic facial emotion recognition rate varies with *L*, *R*

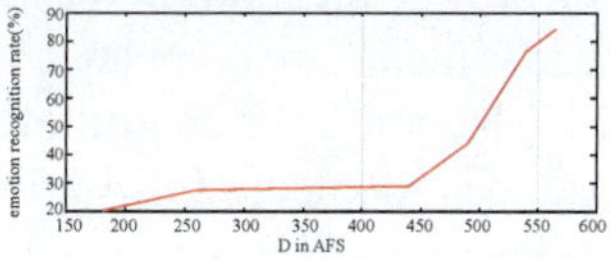

Fig. 4.4 Dynamic facial emotion recognition error rates varies with *D*

4.6.2 Simulations and Analysis

L, *R* and *D* are the three parameters included in the LRS-based AFS, where *L* is the number of features added per cycle, *R* is the number of features removed per cycle, and *D* is the total number of features left by the final cycle. When the parameters *L*, *R* and *D* take different values, the recognition rate of dynamic facial emotion recognition changes with each parameter is shown below.

Fig. 4.5 Dynamic facial emotion recognition rate as a function of k

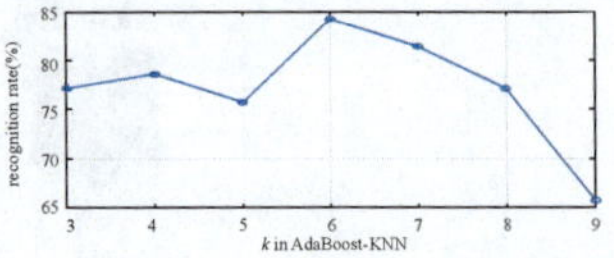

Fig. 4.6 Dynamic facial emotion recognition error rates as a function of i

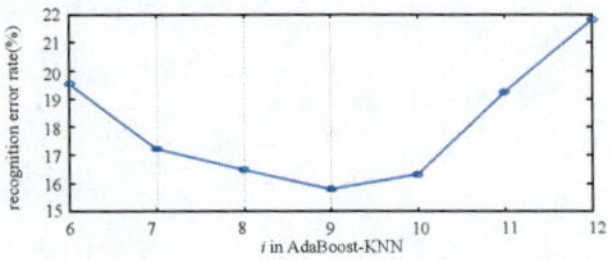

Table 4.1 Dynamic facial emotion recognition rate for selected algorithms

Algorithm	Recognition rate (%)	Variance	Kappa coefficient
AdaBoost-KNN	81.42	7.38	0.78
AFS-AdaBoost-KNN	88.57	3.41	0.86
AdaBoost-KNN-DO	91.42	5.06	0.89
AFS-AdaBoost-KNN-DO	94.28	0.67	0.93

When $D = 540$, $R = 124$, L is changed, and when $D = 540$, $L = 180$, R is changed, the change of dynamic facial emotion recognition rate is shown in Fig. 4.3. When $L = 125$, $R = 124$, the number of feature groups selected last is changed, the change of dynamic facial emotion recognition rate is shown in Fig. 4.4.

In the convergence judgment experiment of recognition error rate, with the increase of k, the recognition accuracy first goes up and then goes down, as shown in Fig. 4.5. With the increase of the number of iterations (i), the recognition error first decreases and then increases. Variation trends are shown in Fig. 4.6.

In order to discuss the effect of LRS algorithm and AdaBoost-KNN optimization algorithm on dynamic emotion recognition rate, we designed a comparative experiment. The recognition rates, variances and kappa coefficients of the three groups of experiments are compared, as shown in Table 4.1.

The confusion matrix of Adaboost-KNN is shown in Fig. 4.7. When AFS algorithm is adopted(AFS-AdaBoost-KNN), the recognition results are shown in Fig. 4.8. When the direct optimization algorithm of AdaBoost-KNN (AdaBoost-KNN-DO) is adopted, the identification result confusion matrix is shown in Fig. 4.9. When both AFS and the direct optimization algorithm of AdaBoost-KNN convergence judgment(AFS-AdaBoost-KNN-DO) are adopted, the confusion matrix is shown in Fig. 4.10. Kappa coefficient is a measure of classification accuracy. In general, kappa coefficient falls between 0 and 1 and can be divided into five groups to represent different levels of consistency. 0.0–0.20 represents a slight agreement, 0.21–0.40 represents a fair agreement, 0.41–0.60 represents a moderate agreement, 0.61–0.80 represents a high agreement and 0.81–1 represents almost perfect agreement. The result shows that the recognition result of method Adaboost-KNN receives a high agreement.

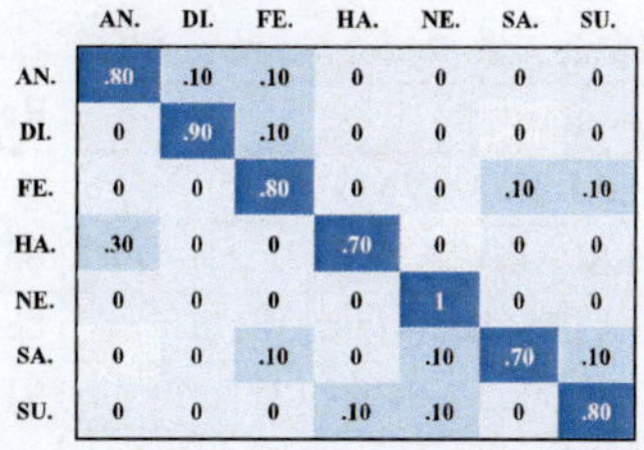

Fig. 4.7 Confusion matrix of emotion recognition by using AdaBoost-KNN

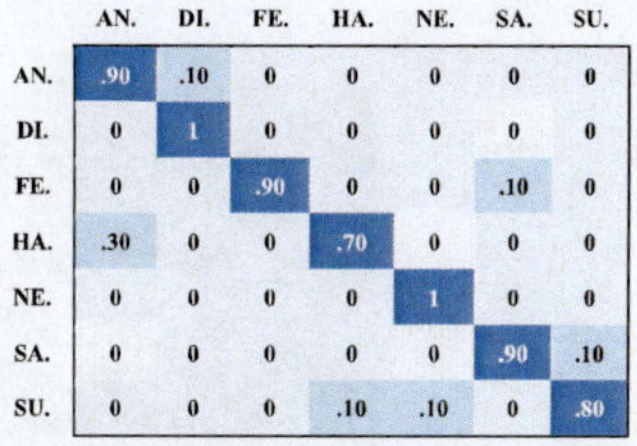

Fig. 4.8 Confusion matrix of emotion recognition using AFS-AdaBoost-KNN

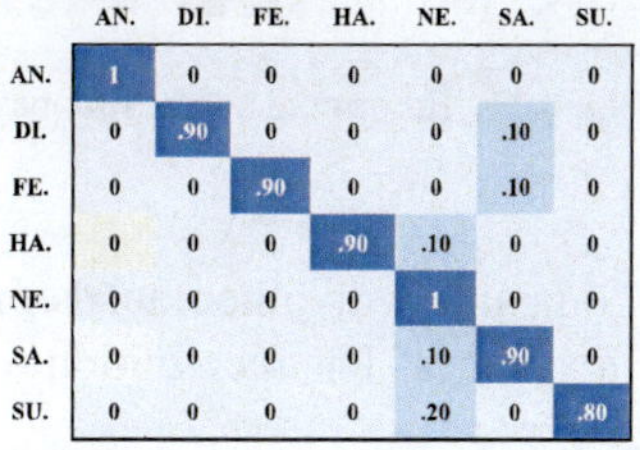

Fig. 4.9 Confusion matrix of emotion recognition by using AdaBoost-KNN-DO

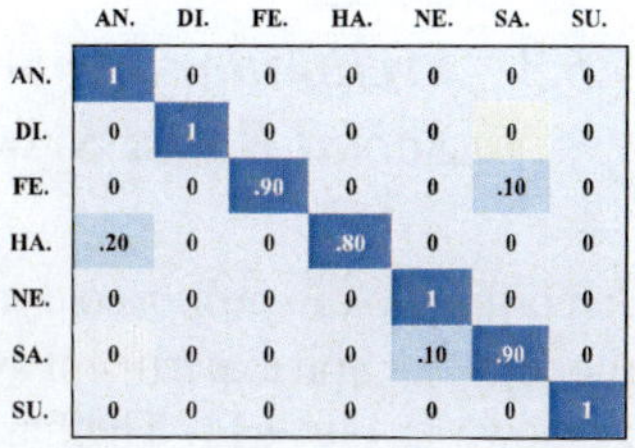

Fig. 4.10 Confusion matrix of recognition using AFS-AdaBoost-KNN-DO

When using AFS-AdaBoost-KNN, AdaBoost-KNN-DO and AFS-AdaBoost-KNN-DO, the recognition result receives almost perfect agreement. When $n = 3$ and $k = 4$, the progressive significance of Friedman test [33] is less than 0.01, therefore it can be concluded that there are significant differences between the results produced by the four methods.

The experimental results show that when the L is 125, R is 124, and the number of features left behind is 566, the facial emotion recognition algorithm achieves the highest emotion recognition rate 94.28%, 5.71% higher than that of single KNN. With the improvement of the method, the variance of the experimental results also decreases to varying degrees, and the variance of the proposal is reduced to 0.67.

Table 4.2 Recognition results obtained for different methods

Algorithm	Recognition rate (%)	Variance
GM-PSO-SVM [15]	89.00	\
LSFA-SVM [17]	85.95	3.82
ODFA-AdaBoost [18]	88.78	\
Chaotic FA-KNN [23]	91.45	\
AFS-AdaBoost-KNN-DO	94.28	0.67

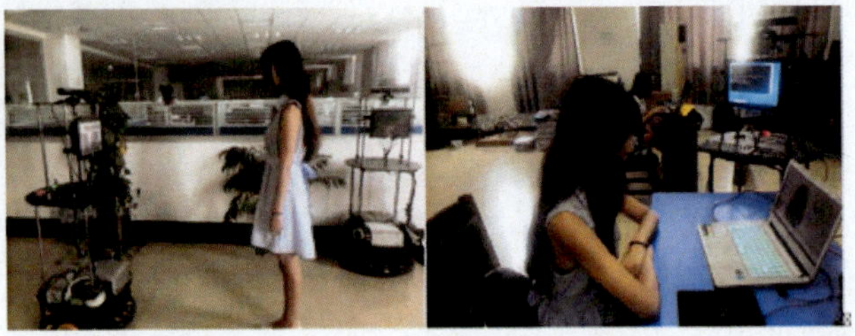

Fig. 4.11 Emotional social robot system

Compared with other facial emotion recognition methods the proposal exhibits great advantages. The recognition results of different methods are shown in Table 4.2.

4.6.3 Preliminary Application Experiments on Emotional Social Robot System

According to the proposed dynamic facial emotion recognition algorithm, a human-robot interaction experiment system is built. The environment required by the system is shown in Fig. 4.11. The hardware of facial expression interaction system mainly includes two parts, which are Kinect sensor used to obtain facial expression and wheeled robot used to express and interact emotions.

Dynamic emotion recognition is done in the preliminary application. Ten volunteers aged 18–28 are invited to build up the experimental data, and they are postgraduates of our laboratory or undergraduate students. Each volunteer expresses seven primary emotions: happiness, anger, sadness, disgust, surprise, fear, and neutral. The data samples of one volunteer are collected as shown in Fig. 4.12.

According to the real-time dynamic facial expression of humans, 70 dynamic emotions of 10 volunteers are recognized by using the feature point matching and the introduced Candide-3-based dynamic feature points matching. Experiment results are shown in Table IV, and the confusion matrix of emotion recognition is shown in Fig. 4.13.

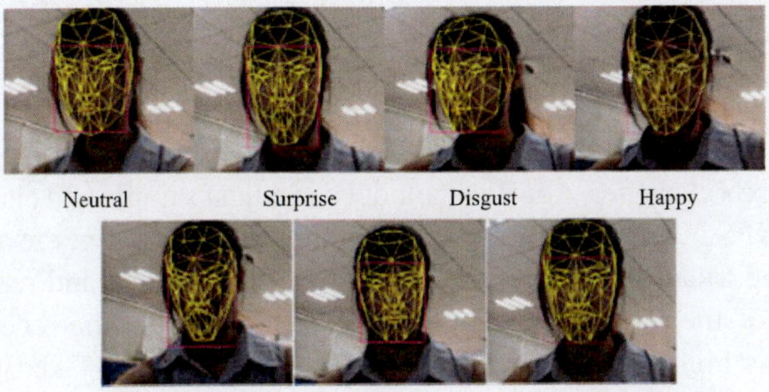

Fig. 4.12 The data samples of one volunteer

	AN.	DI.	FE.	HA.	NE.	SA.	SU.
AN.	.80	.10	.10	0	0	0	0
DI.	0	.80	.10	0	0	.10	0
FE.	0	0	.70	0	0	0	.30
HA.	0	0	.10	.90	0	0	0
NE.	0	0	.10	0	.90	0	0
SA.	.10	0	0	0	0	.90	0
SU.	0	0	.30	0	0	0	.70

Fig. 4.13 Confusion matrix of recognition results obtained by using AFS-AdaBoost-KNN-DO

According to the result analyses of average recognition rate and confusion matrices, the average recognition rate is 81.42% and the kappa coefficient is 0.78 by using the proposal. Recognition results produced by the method AFS-AdaBoost-KNN-DO are substantially consistent with the test sample. While the average recognition rate was 72.34% when AFS was not used. The application experiment of human-robot interaction system shows that the recognition system has a high recognition rate of five emotions: calm, sad, happy, disgusted and angry. But fear and surprise are not easily recognized. The expressions of fear and surprise emotion all contain similar characteristics such as binocular round stare, so it is difficult to distinguish between them.

The results show that the dynamic facial emotion recognition algorithm proposed in this chapter has good real-time performance and good recognition effect in man-machine interaction system. Moreover, the proposal takes into account the contribution of different feature points to facial expression recognition, which can find out the variation of main feature points in facial expression, in such a way that emotion recognition rate was improved by 9.08%.

4.7 Summary

AFS-AdaBoost-KNN-DO is proposed for dynamic facial emotion recognition. On the basis of KNN classifier, AdaBoost model can adjust sample and classifier weights. Meanwhile, AFS feature selection method is adopted. Moreover, the direct optimization is adopted to automatically select the model parameters and realize the global optimization of the model. Experimental results show that the recognition rate of the proposal is higher than that of the AdaBoost-KNN, AFS-AdaBoost-KNN and AdaBoost-KNN-DO. It is also higher than other traditional recognition methods, such as AdaBoost, KNN, SVM, etc.

Emotion recognition and information understanding are getting more and more popular in human-robot interaction, robots recognize, understand, and adapt human emotions through behavior would be an interesting research topics. The preliminary application experiments via our emotional social robot system giving delivers the strong evidence that the proposed AFS-AdaBoost-KNN-DO is qualified to achieve dynamic facial emotion recognition, and indicates the dynamic emotion understanding ability of robots in human-robot interaction.

References

1. K. Anderson, P.W. Mcowan, A real-time automated system for the recognition of human facial expressions. IEEE Trans. Syst. Man Cyber. Part B: Cybern. **36**(1), 96–105 (2016)
2. Z.F. Wang, L. Peyrodie, H. Cao, Slow walking model for children with multiple disabilities via an application of humanoid robot. Mech. Syst. Signal Process. **68**, 608–619 (2016)
3. C. Navarretta, Predicting emotions in facial expressions from the annotations in naturally occurring first encounters. Knowl.-Based Syst. **71**, 34–40 (2014)
4. E. Vezzetti, F. Marcolin, G. Fracastoro, 3D face recognition: an automatic strategy based on geometrical descriptors and landmarks. Auton. Syst. **62**(12), 1768–1776 (2014)
5. L.F. Chen, M. Wu, M.T. Zhou, Zhentao Liu, J.H. She, K. Hirota, Dynamic emotion understanding in human-robot interaction based on two-layer fuzzy SVR-TS model. IEEE Trans. Syst. Man Cyber.: Syst. **50**(2), 490–501 (2020)
6. L.F. Chen, Z.T. Liu, M. Wu, M. Ding, F.Y. Dong, K. Hirota, Emotion-age-gender-nationality based intention understanding in human-robot interaction using two-layer fuzzy support vector regression. Int. J. Soc. Robot. **7**(5), 709–729 (2015)
7. Y.S. Dong, H. Wu, X.K. Li, C.Q. Zhou, Q.T. Wu, Multi-scale symmetric dense micro-block difference for texture classification. IEEE Trans. Circuits Syst. Video Technol. **29**(12), 3583–3594 (2018)
8. C. Zhou, L. Wang, Q. Zhang, X. Wei, Face recognition based on PCA and logistic regression analysis. Int. J. Light Electron Optics **125**(20), 5916–5919 (2014)
9. L. F. Chen, M. Wu, M. T. Zhou, J. H. She, F. Y. Dong, K. Hirota. Information-driven multi-robot behavior adaptation to emotional intention in human-robot interaction. IEEE Trans. Cognit. Dev. Syst. **10**(3), 647–658 (2018)
10. D. Smeets, P. Claes, J. Hermans, D. Vandermeulen, P. Suetens, A comparative study of 3-D face recognition under expression variations. IEEE Trans.Syst. Man Cybern.: Part C **42**(5), 710–727 (2012)
11. E. Vezzetti, F. Marcolin, G. Fracastoro, 3D face recognition: an automatic strategy based on geometrical descriptors and landmarks. Robot. Auton. Syst. **62**(12), 1768–1776 (2014)

12. A.H. Matamoros, A. Bonarini, E.E. Hernandez, Facial expression recognition with automatic segmentation of face regions using a fuzzy based classification approach. Knowl.-Based Syst. **110**, 1–14 (2016)
13. L. Zhang, K. Mistry, S.C. Neoh, C.P. Lim, Intelligent facial emotion recognition using moth-firefly optimization. Knowl.-Based Syst. **111**(1), 248–267 (2016)
14. B. Xue, M. Zhang, W.N. Browne, Particle swarm optimization for feature selection in classification: a multi-objective approach. IEEE Trans. Syst. Man Cybern. **43**(6), 1656–1671 (2013)
15. Y. Zhang, L. Zhang, S.C. Neoh, K. Mistry, A. Hossain, Intelligent affect regression for bodily expressions using hybrid particle swarm optimization and adaptive ensembles. Expert Syst. Appl. **42**(22), 8678–8697 (2015)
16. M. Alweshah, S. Abdullah, Hybridizing firefly algorithms with a probabilistic neural network for solving classification problems. Appl. Soft Comput. **35**, 513–524 (2015)
17. O.P. Verma, D. Aggarwal, T. Patodi, Opposition and Dimensional based modified firefly algorithm. Expert Syst. Appl. **44**, 168–176 (2016)
18. Y. Guo, Y. Liu, A. Oerlemans, Deep learning for visual understanding: a review. Neurocomputing **187**, 27–48 (2016)
19. J.C. Liu, L. Zhang, X. Chen, J.W. Niu, Facial landmark automatic identification from three dimensional(3D) data by using Hidden Markov Model(HMM). Int. J. Ind. Ergon. **57**, 10–22 (2017)
20. L.F. Chen, M.T. Zhou, W.J. Su, M. Wu, J.H. She, K. Hirota, Softmax regression based deep sparse autoencoder network for facial emotion recognition in human-robot interaction. Inf. Sci. **428**, 49–61 (2018)
21. Z.H. Yu, C. Li, Texture classification and retrieval using shearlets and linear regression. IEEE Trans. Cybern. **45**(3), 358–369 (2015)
22. J. Wang, R. Xiong, J. Chu, Facial feature points detecting based on Gaussian Mixture Models. Pattern Recogn. Lett. **53**, 62–68 (2015)
23. W. Zhang, Y. Zhang, L. Ma, J.W. Guan, S.J. Gong, Multimodal learning for facial expression recognition. Pattern Recogn. **48**(10), 3191–3202 (2015)
24. S. Elaiwat, M. Bennamoun F. Boussaid, A curvelet-based approach for textured 3D face recognition. Pattern Recogn. **48**(4), 1235–1246 (2015)
25. X.L. Li, G.S. Cui, Y.S. Dong, Graph regularized non-negative low-rank matrix factorization for image clustering. IEEE Trans. Cybern. **47**(71), 3840–3853 (2017)
26. L. F. Chen, M. Wu, J. H. She, F. Y. Dong, K. Hirota, Three-layer weighted fuzzy SVR for emotional intention understanding in human-robot interaction. IEEE Trans. Fuzzy Syst. **26**(5), 2524–2538 (2018)
27. N. Deng, Y.B. Pei, Z.G. Xu, Face recognition with single sample based on Candide-3 reconstruction model, in *Applied Mechanics and Materials*, pp. 3623–3628 (2013)
28. X.L. Li, Q.M. Lu, Y.S. Dong, A manifold regularized set-covering method for data partitioning. IEEE Trans. Neural Netw. Learn. Syst. **29**(5), 1760–1773 (2018)
29. Q.M. Lu, X. Li, Y.S. Dong, Structure preserving unsupervised feature selection. Neurocomputing **301**, 36–45 (2018)
30. J. Maillo, F. Ramírez, I. Triguero, F. Herrera, K-NN-IS: An iterative spark-based design of the k-nearest neighbors classifier for big data. Knowl.-Based Syst. **117**, 3–15 (2017)
31. Y. Wen, L. He, P. Shi, Face recognition using difference vector plus KPCA. Dig. Signal Process. **22**(1), 140–146 (2012)
32. The Japanese Female Facial Expression (JAFFE) Database, http://www.kasrl.org/jaff.html
33. J. Demišar, D. Schuurmans, Statistical comparisons of classifiers over multiple data sets. J. Mach. Learn. Res. **7**(1), 1–30 (2006)

Chapter 5
Weight-Adapted Convolution Neural Network for Facial Expression Recognition

The weight-adapted convolution neural network (WACNN) is proposed to extract discriminative expression representations for recognizing facial expression. It aims to make good use of the convolution neural network's potential performance in avoiding local optimal and speeding up convergence by hybrid genetic algorithm (HGA) with optimal initial population, in such a way that it realizes deep and global emotion understanding in human-robot interaction. Moreover, the idea of novelty search is introduced to solve the deception problem in the HGA, which can expend the search space to help genetic algorithm jump out of local optimum and optimize large-scale parameters. In the proposal, the facial expression image preprocessing is conducted first, then the low-level expression features are extracted by using principal component analysis. Finally, the high-level expression semantic features are extracted and recognized by WACNN which is optimized by HGA.

5.1 Introduction

Facial expression conveys abundant information about emotions [1], intentions and other internal states [2], which is one of the most natural ways for nonverbal communications between people [3]. Therefore, acquiring facial expression information is very important for emotion understanding. In order to understand the emotion, it is necessary to obtain the expression information first by recognizing facial expression. Moreover, facial expression recognition (FER) is propitious to recognize human emotions by robots and promote the development of artificial intelligence. In addition, FER has gained widespread attention recently [4], since its widely range of applications [5], e.g., fatigue driving test, human behaviour recognition, human-robot interaction (HRI), etc.

Although much attention is given on FER in the past few decades, FER continues to be a challenging task for extracting representative expression features which are

L. Chen et al., *Emotion Recognition and Understanding for Emotional Human-Robot Interaction Systems*, Studies in Computational Intelligence 926,
https://doi.org/10.1007/978-3-030-61577-2_5

discriminative of imperceptible changes in facial expression [6–8]. In recent years, deep learning (DL) has attracted many researchers, due to that numerous breakthrough have been observed in some fields by using DL method [9], e.g., speech recognition, image detection, image recognition and so on. Deep neural networks (DNNs) don't need handcrafted feature extraction techniques, and can still learn meaningful representations from the raw data [10]. The representative network structure in DL are autoencoders (AEs), convolutional neural networks (CNNs), recurrent neural networks, and so on [11]. Since that the CNNs are established by the visual principles of organisms, CNNs have excellent performance in image processing [12]. The principal component analysis (PCA) can efficiently reduce the correlation between features, which is similar to the process of retina processing image (the retina performs a decorrelation operation to get a lower redundant representation of the input image and transmit it to the brain). Moreover, according to the research about recognizing emotion from faces by human, in the first stage, the extraction of low-level prominent expression feature is happening in occipital temporal cortex; at other stages, the learning of high-level expression feature is occurring in other brain areas, the emotion perception as well [13]. Inspired by this working mechanism on facial expression perception of human, for the sake of dealing with such challenges to produce discriminative expression representations, we develop a novel FER system which exploits PCA to extract low-level expression features, and the deep CNN (DCNN) is designed for learning high-level expression semantic feature. Furthermore, due to that the PCA and CNNs are both motivated by visual principles, the low-level expression features extracted by PCA can fit the concept of CNNs well, and PCA also can facilitate the extraction of high-level expression semantic features in subsequent steps by using DCNN.

The weights of DNN are the key to determine the representational performance of it, so it is obvious that the optimization algorithm of weights is vital for DNN. Generally, the optimization algorithms used in DNN are gradient-based algorithms which calculate or approximate gradients and optimize weights via gradient descent/ascent. Nevertheless, gradient-based approaches tend to trap into poor local optimal solution. The evolutionary algorithm (EA) is a kind of population-based algorithm, has shown powerful global search capability [14]. In virtue of locally optima insensitive and gradient-free properties, some researchers use the EA to optimize the weights or structure of neural networks [15, 16]. Among various EAs, genetic algorithm (GA) which operates directly on structural objects without the limitation of function's derivation is one of the most popular algorithms. And GA can automatically optimize search space and adaptively adjust the search direction. In light of these properties, to get the best performance of the DCNN for the discriminative representations, a hybrid GA (HGA) is developed to adaptively optimize the weights of DCNN. GA suffers when optimizing large-scale parameters in a relative short time [16], which is more suitable for optimizing small-scale parameters. However, DCNN has tens of thousands of weights needed to be optimized, it is difficult to converge using GA. Therefore, the weights optimized by accelerated gradient descent (AGD) will be taken as a chromosome of GA's initial population to accelerate the convergence of the GA. In general, GA exhibitis deception problem, which may actively misdirect search toward dead ends, especially when there is the super-individual. The super-individual will make GA lose the ability to explore new space, and easy to converge to local optima. Consequently, the idea of novelty search (NS) is introduced [17] to expand the search space.

In this chapter, to extract discriminative facial expression features for FER, a weight-adapted convolution neural network (WACNN) by using HGA is proposed. To this end, the main contributions are summarized as follows:

1. Inspired by the working mechanism on facial expression perception of human, the WACNN is proposed that exploits PCA to extract low-level expression features, and the DCNN is utilized to learning high-level expression semantic features. Moreover, weights are vital for the performance of DCNN, a HGA is applied to adaptively optimize the weights of DCNN.
2. The weights optimized by AGD will be taken as an initial chromosome of HGA, so that it can speed up the convergence of HGA, and also can be used as a prior knowledge.
3. To solve the problem of deception, the idea of NS is introduced. The NS is used to find the most novel individual in the high adaptable individuals, so that it is able to extend the search space, as well as enhance the global search capability of HGA.

5.2 Facial Expression Image Preprocessing

In general, in the facial expression image, besides the key facial expression information, it also contains the noise of the background, hair, and so on, which will interfere with the recognition of the facial features. Thus, before extracting the expression feature, it is necessary to do preprocess.

The geometry normalization is conducted first to extract key face areas by the face detection algorithm of Viola–Jones [46], so that the richer regions of expression distribution are obtained refer to Fig. 5.1. Then, the gray normalization is applied by histogram equalization, so that the influence of light intensity on expression features is reduced.

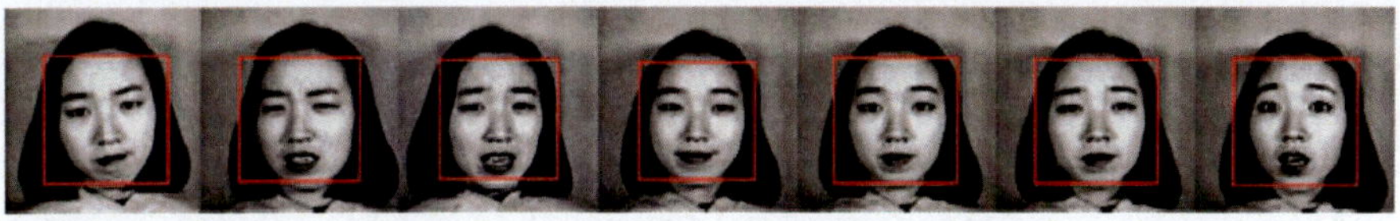

Fig. 5.1 Geometric normalization

5.3 Principal Component Analysis for Extracting Expression Feature

Because there is strong correlation between adjacent pixels in the facial expression image, the image data exhibit a great deal of redundancy. So, in order to reduce correlation between features and get the low-level expression features, the PCA is applied to the image data. And the process of it is as follows:

First, normalize each facial expression data's intensity by mean normalization as follows,

$$a^{(j)} = a^{(j)} - \frac{1}{n}\sum_{j=1}^{n} a^{(j)} \tag{5.1}$$

Then, calculate a's covariance matrix and obtain eigenvector U, then a is represented by a base vector $\{u_1, u_2, ..., u_n\}$ of U, and the a_{rot} is given by

$$a_{rot} = U^T a \tag{5.2}$$

Next, to make each of features have unit variance, rescale the data by

$$a_p = \frac{a_{rot}}{\sqrt{\lambda + \varepsilon}} \tag{5.3}$$

where λ is the eigenvalue of covariance matrix, ε is regularization factor.

Finally, we perform ZCA (Zero-phase Component Analysis) whitening to make the covariance matrix into identity matrix by

$$a_z = U a_p \tag{5.4}$$

After applying PCA, the low-level expression features are formed. The low-level features' redundancy are greatly reduced and the edge of the expression images are enhanced.

5.4 Weight-Adapted Convolution Neural Network for Recognizing Expression Feature

The concrete structure of DCNN in WACNN is shown in Fig. 5.2. Owing to the difference between various facial expression is mainly in eyebrows, eyes, mouth, etc., the changes of other parts are limited, we use a relative large convolution kernel to make it have a larger receptive field that can detect significant feature changes at each time of its movement.

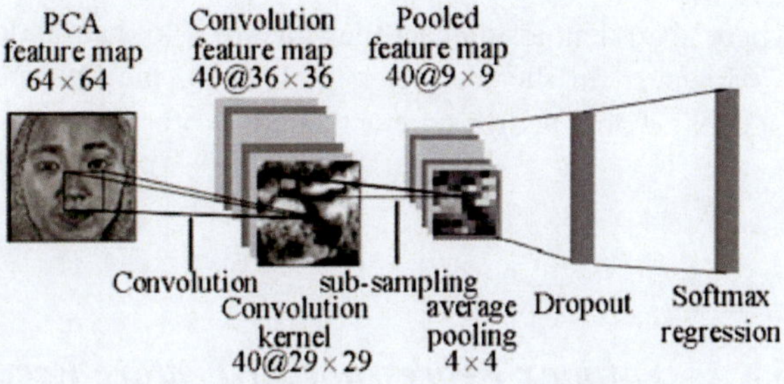

Fig. 5.2 Structure of DCNN in WACNN

5.4.1 *Feature Learning Based on Deep Convolution Neural Network*

In DCNN, convolution layer is the core of it. The convolution kernel enables DCNN to have weight sharing and local receptive filed functions. The convolution can be expressed by

$$c_j^{(l+1)} = f\left(\sum_{i=M_j} c_i^l * W_{ij}^{(l)} + b_j\right) \tag{5.5}$$

where $*$ is the convolution operation, $c_i^{(l)}$ denotes the *i-th* expression feature data in the *l-th* layer of DCNN, $W_j^{(l)}$ is the *j-th* convolution kernel, and b_j denotes the corresponding bias. The $f(\cdot)$ denotes a sigmoid function. As shown in Fig. 5.2, in the convolution layer, the WACNN sets 40 convolution kernels with kernel size of 29×29 , and the stride is 1.

The average pooling of sub-sampling also is the major component of DCNN. On the one hand, it reduces the feature map and simplifies the complexity of network computing; on the other hand, it compresses the features to extract the main features. The sub-sampling is implemented by

$$c_j^{(l+1)} = f\left(c_j^{(l)} * \frac{1}{h^2}\right) \tag{5.6}$$

where $c_j^{(l)}$ denotes *j-th* feature map of *l-th* layer, and h denotes dimension of pooling. As shown in Fig. 5.2, in sub-sampling layer, the sub-sampling dimension is 4, and the stride is also set to 4.

The dropout layer is introduced [21] to handle the overfitting problem in DCNN. In the training stage of DCNN, the operation of dropout lets the values of some neurons in Dropout layer be are set to zero randomly at each iteration, but their real values of the neurons will be preserved. This means that the effective structure of DCNN is not same at each iteration, the calculation process is given as

$$d = Z(p)d \tag{5.7}$$

where $Z(p)$ denotes that set the value of d to 0 at random with the probability p.

However, in the test stage of DCNN, the DCNN becomes completed where DCNN's whole neurons are activated, expressed as

$$d = (1 - p)d \tag{5.8}$$

5.4.2 *Softmax Regression for Feature Recognition*

Softmax regression is applied to recognize the learned feature of DCNN. Given a test input z, the hypothesis function $h_{\vartheta}(z)$ is used to estimate the probability of each of the possible values of the q class labels. Thus, hypothesis function outputs q estimated probabilities, and the elements of q sum up to 1. $h_{\vartheta}(z)$ is given by

$$h_{\vartheta}(z^{(i)}) = \begin{bmatrix} p(y^{(i)} = 1|z^{(i)}; \vartheta) \\ p(y^{(i)} = 2|z^{(i)}; \vartheta) \\ \vdots \\ p(y^{(i)} = q|z^{(i)}; \vartheta) \end{bmatrix} = \frac{1}{\sum_{j=1}^{q} e^{\theta_j^T z(i)}} \begin{bmatrix} e^{\vartheta_1^T z(i)} \\ e^{\vartheta_2^T z(i)} \\ \vdots \\ e^{\vartheta_q^T z(i)} \end{bmatrix} \tag{5.9}$$

where $y^{(i)} \in \{1, 2, ..., q\}$ is the labels of training set.

The Softmax regression's cost function is given by

$$J(\vartheta) = -\frac{1}{m}\left[\sum_{i=}^{m}\sum_{j=1}^{q} 1\{y^{(i)} = j\} \log \frac{e^{\vartheta_j^T z^{(i)}}}{\sum_{l=1}^{q} e^{\vartheta_j^T z^{(i)}}}\right] + \frac{\lambda}{2}\sum_{i=1}^{q}\sum_{j=0}^{n} \vartheta_{ij}^2 \tag{5.10}$$

where $1\{\cdot\}$ is the indicator function, and $\frac{\lambda}{2}\sum_{i=1}^{q}\sum_{j=0}^{n} \vartheta_{ij}^2$ is a weight decay term.

5.5 Hybrid Genetic Algorithm for Optimizing Weight Adaptively

GA searches for the optimal solution by simulating the natural selection and genetic mechanism in Darwin's theory of biological evolution [47], which has better global search capability. In light of this, the HGA is proposed to optimize the weights of WACNN adaptively for better representative facial expression features, whose implementation steps are summarized as follows:

Step 1: Obtain optimized weights θ_1 by optimizing DCNN using AGD.

Step 2: Initialize the population P with N chromosomes, one of which is the weights obtained by AGD $P_N = \theta_1$.

Step 3: Randomly shuffle training set, divide the training set into mb mini-batches.

Step 4: Evaluate the fitness of each chromosome $F = f(P)$.

Step 5: If there are $F = 0$, then go to *Step 6*, else go to *Step 8*.

Step 6: Insert the chromosome with a fitness of 0 into a fixed archive.

Step 7: If the number of chromosomes in the fixed archive is higher than K, then calculate the average distance D by (5.15) and sort P with ascending order by D where the fitness of P is 0, else go to *Step 8*.

Step 8: Sort P with descending order by F.

Step 9: Perform genetic operations of selection, crossover, and mutation to produce a new population, add the chromosome with the best fitness values or the chromosome with the largest D directly to the new population.

Step 10: Repeat *Step 2* $\sim$ *Step 9* until the maximum number of generations is reached, and the best chromosome is obtained.

As shown in *Step 2*, the weights optimized by AGD will be taken as a chromosome of HGA's initial population in the proposal, so that it can accelerate the convergence of the algorithm and also can be used as a priori knowledge. To preserve the diversity of the population, the remaining chromosomes are obtained by random initialization. In the chromosome representations of GA, the real-coded method is applied in the proposal.

AGD update is given by

$$v_t = \gamma v_{t-1} + \alpha \nabla_\theta J(\theta - \gamma v_{t-1}; x^{(i)}, y^{(i)}) \tag{5.11}$$

$$\theta_t = \theta_{t-1} - v_t \tag{5.12}$$

where $\nabla_\theta J(\cdot)$ is the gradient of (5.10), θ stands for all weights of DCNN, $\gamma \in (0, 1]$, α is a learning rate, and the v denotes a velocity vector.

At each genetic iteration, the fitness evaluation is conducted first that is defined as (5.13), which is the error of current training set.

$$\text{fitness} = 1 - Accuracy \tag{5.13}$$

Then, population is arranged in descending sort on the basis of the value of fitness function. And the operation of selection by using stochastic uniform selection is performed, that is, according to the value of an 1-by-N matrix to select corresponding chromosomes, which are random integers generated by a uniform distribution. Next, the genetic operators of crossover and mutation are implemented by a two-point crossover, and applying additive Gaussian noise to the population as

$$P^{g+1} = P^g + \sigma \kappa \tag{5.14}$$

where σ is the scaling factor, and $\kappa \sim N(0,I)$, Gaussian distribution. However, the chromosomes with the best fitness will not perform this genetic operators. They will be directly passed on to the next population, which is called elite strategy.

In general, GA has the deception problem, which may actively misdirect search toward dead ends, especially when there is the super-individual. Due to that the weights optimized by AGD will be taken as a chromosome of HGA's initial population in the proposal, after a certain number of crossover, most chromosomes will have a similar fitness, which are all around 0. This will enhance the local search ability of HGA, seriously weaken the ability of global search, and make it difficult to jump out of local optimum. In this case, it is hard for HGA to further select better chromosomes. To solve this, the the idea of NS is introduced to expend the search space.

In NS, it seeks novelty of behavior. NS uses a novelty metric, instead of the fitness function, which rewards behaving differently from prior chromosomes in the search. Moreover, NS adds an archive that remembers chromosomes that were highly novel when they were first discovered. However, in the proposal, the fitness function is still used to reward progress towards objective, and the novelty metric is as the second evaluation indicator. When the fitness of the population is no longer improved, in the implementation of the elite strategy, in addition to the most adaptable individual will be directly inherited to the next generation, and the most novel individual in the high adaptable individuals will also be directly inherited to the next generation.

The novelty metric given by

$$D = \frac{1}{d}\sum_{j=0}^{d} dist(P, A_j) \tag{5.15}$$

where the P is chromosomes of current population whose fitness is 0, A is a fixed archive which stores the latest $2N$ chromosomes with a fitness of 0, where A_j is the jth-nearest neighbor of P with respect to metric Euclidean distance. If the average distance D is big, it's in a sparse region, otherwise, it's in a dense area. The chromosome with the largest average distance will be the most novelty chromosome, which will be directly passed on to the next population.

In the meanwhile, to improve the optimization efficiency of HGA, the stochastic mini-batch is applied in the HGA. Mini-batch calculations also reduce the use of computing resources. Moreover, if training sets are given in a meaningful order, this may misdirect search and lead to poor convergence. Therefore, in order to avoid this situation, the data of the training set will be randomly shuffled before each epoch.

5.6 Experiments

There are acknowledged six basic expressions, namely, angry, disgust, fear, happiness, surprise and sadness [48]. In the experiments, seven facial expression are considered, that is, except the six basic expression above, the neutral is also included.

Fig. 5.3 Some expression image samples in CK+

Fig. 5.4 Some expression image samples in JAFFE

Fig. 5.5 Some expression image samples in SFEW 2.0

Moreover, k-fold cross validation method is applied to verify the algorithm, namely, the data set will be evenly divided into k parts, and in these databases, k are five. And for each test, four of them are training sets and the remaining one is a test set. Moreover, in the experiments, the γ and α in (5.11) are set to 0.95 and 0.1, and σ in (5.14) is set to 0.005.

The CK+ [49] database contains 210 subjects with different race and gender. It's composed of sequences of expression, and the neutral state and the apex state frames are chosen, so, there totally 500 frames of the seven basic expressions. The JAFFE [50] database has 213-grayscale images with ten girls, and each girl has two to four images of the seven facial expression. The static facial expressions in the wild 2.0 (SFEW 2.0) [51] database includes facial expression data that close to real world environment which are extracted from movies. And this database includes unconstrained facial expressions, different head postures, the large age range, occlusion, different focal points, different facial partial resolutions and near-real-world lighting. The data are divided into three parts, test set, training set, and validation set. Due to that the labels of test set can not access for the public, the test set and training set are used in the experiments. In total, 1218 images are used in the experiments which cover the seven basic expressions. Figures 5.3, 5.4, and 5.5 and show some expression image samples in these database.

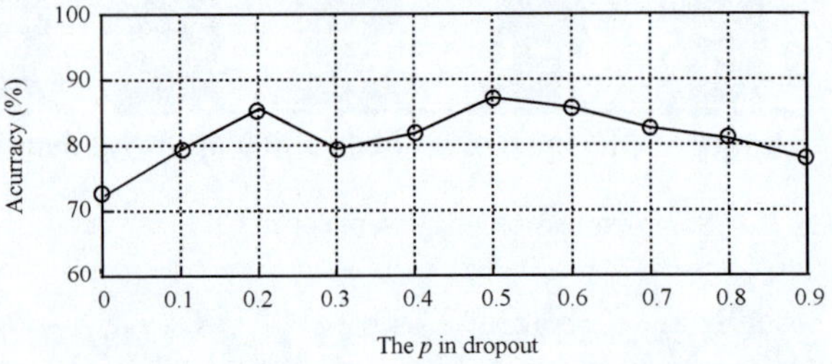

Fig. 5.6 Influence of parameter p on the accuracy

In DL, a common problems is that it needs lots data to learn in training stage, or the model's performance will be poor. However, the benchmark databases does not meet this requirement in general. Thus, to address this problem, the original images in the benchmark databases are symmetrically transformed, so that it doubles the size of the training set. Two experiments on JAFFE database are carried out: one uses the enriched data, the another only uses the original data. Moreover, the p in dropout is 0 in the experiments, and the AGD is used to optimized the DCNN. The experimental result is that accuracy of the original data is 80.85% and accuracy of the enriched data is 87.23%, showing that the enrich data can improve performance of DCNN effectively.

Figure 5.6 shows that the accuracy varies with the value of p in the dropout layer when AGD is used to optimize the DCNN. It can be seen that the accuracy has an increasing trend, when p is increasing. According to the Fig. 5.6, we choose $p = 0.5$ as the optimal value.

To explain the role of PCA in the proposal, the comparison of before and after expression feature extraction by PCA is given in Fig. 5.7. As shown there, the PCA can basically reconstruct the corresponding original high-dimensional vectors by means of low-dimensional representation vectors and eigenvector matrices. Although part of the image information is lost, it will not affect the image quality; some edges of facial features in the image are strengthened compared with the image before expression feature extraction by PCA, and this kind of low-level expression features facilitates the extraction of high-level expression semantic features in subsequent steps by DCNN. The comparison experiments without PCA (WPCA) is conducted and the experimental results are given in Table 5.1. The experiments with PCA achieve higher accuracy than that of WPCA, so that the extraction of low-level expression features by using PCA is an important part of WACNN.

Experiments on JAFFE, CK+, and SFEW 2.0 databases are conducted to validate the usefulness of the WACNN, and AGD and HGA are used to optimized the DCNN respectively. Furthermore, the training set increases the enriched data. The experimental results on these three benchmark databases are presented in Table 5.1 where the k denotes that the k-th part in the k-fold cross validation, on JAFFE database, the proposal achieves the average accuracy of 94.01%; on CK+ database, the proposal achieves average accuracy of 91.00%; on SFEW 2.0 database, the proposal achieves average accuracy of 49.02%. And the accuracies are improved by 4% $\sim$ 6% using WACNN compared with using AGD with DCNN. It follows that the pro-

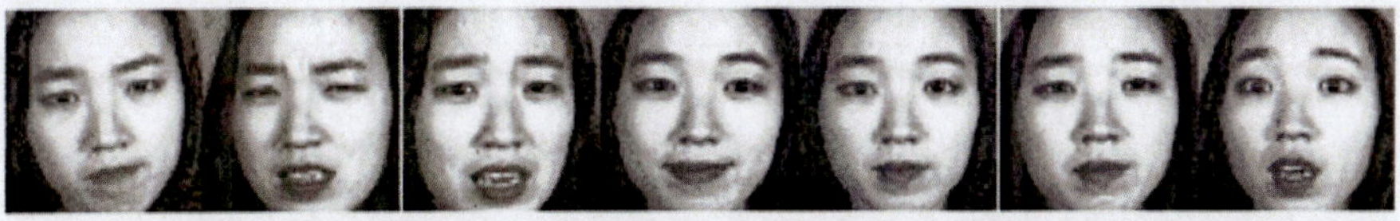

(a) Before expression feature extraction by PCA

(b) After expression feature extraction by PCA

Fig. 5.7 Results of comparative analysis

Table 5.1 Comparisons with other approaches on benchmark databases

k-fold	JAFFE			CK+			SFEW 2.0		
	WACNN	AGD+DCNN	WPCA	WACNN	AGD+DCNN	WPCA	WACNN	AGD+DCNN	WPCA
$k = 1$ (%)	94.60	89.19	67.57	89.11	84.16	70.30	49.39	46.12	30.20
$k = 2$ (%)	97.06	94.12	70.59	93.07	89.11	81.19	47.54	43.85	36.48
$k = 3$ (%)	97.06	94.12	76.47	89.90	84.85	68.69	51.04	46.89	36.93
$k = 4$ (%)	90.91	87.88	69.70	93.94	88.89	75.76	47.33	42.39	30.04
$k = 5$ (%)	90.40	83.87	64.71	89.00	84.00	72.00	49.80	46.53	25.71
Average (%)	94.01	89.84	69.81	91.00	86.20	73.59	49.02	45.16	31.87

posal achieves better performance than that of AGD with DCNN, owing to that the HGA have better global search ability than that of the AGD, which can prevent the HGA from falling into the local optimum and produce better weights for WACNN, so that the WACNN is able to get more discriminative expression representations.

Furthermore, the optimal population is initialized in the HGA, so that the HGA constantly searches for weights with novel behavior and higher adaptability to jump out of the current local optimum based on optimal population in the process of evolution. Figure 5.8 shows the change of cost function value during evolution. As shown in Fig. 5.8, the cost function continues to converge in the process of evolution, and tends to be relatively stable in the end. It should be noted that the cost function curve seems to have a beating after convergence, which is due to that the stochastic mini-batch is applied in the HGA. Thus, the training sample is different in every iterative, which will cause the jitter of the cost function value, but overall the function is heading in the best direction.

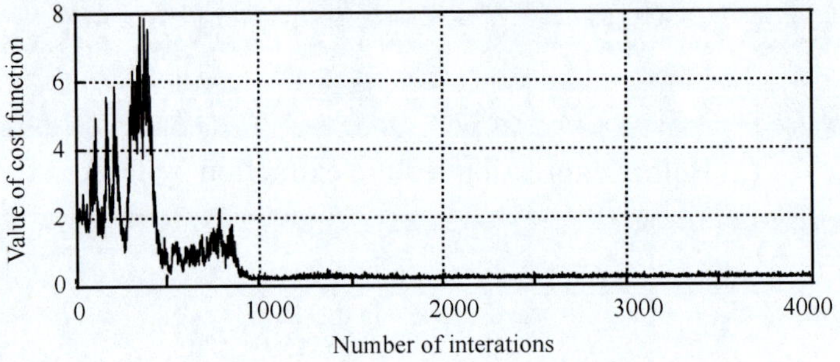

Fig. 5.8 The change of cost function value in the iterative process

Figure 5.9 contains the confusion matrixes of experimental results obtianed on benchmark databases, which are the average accuracies of the five-fold cross validation experiment. In JAFFE database, the recognition result of neutral is 100%; followed are the recognition results of angry, happiness, sadness, surprise and disgust which are more than 90%; and the lowest recognition result is sadness with the accuracy of 80%. However, in CK+ database, the recognition results of surprise is the with the accuracy of 99%; the accuracies of happiness, disgust, sadness, neutral are higher than 89% followed by the recognition result of surprise; the lowest recognition result is fear with the accuracy of 80%; between angry and sadness, fear and sadness, fear and happiness, there is a relatively high confusions. And in SFEW 2.0 database, the accuracies of angry, happiness, neutral, and sadness are higher than 50%, the lowest recognition result is disgust. Owing to that this database is more complex than that of JAFFE and CK+ databases, the accuracies achieved in SFEW 2.0 is lower than them, confusion is also more serious. And between angry and disgust, happiness and disgust, neutral and disgust, neutral and surprise, there is a relatively high level of confusion. These confusion in these databases may be due to the fact that the extracted features can not express small differences in some expressions. It is acknowledged that there are some similarities between facial expressions, because of the small differences in motion features, it may be surprise or fear to stare, disgust or fear to squeeze. In addition, some samples in the benchmark database are ambiguous, and it is difficult for even people to predict the labels of these samples, which may mislead the WACNN to learn bad features, such as the samples shown in Fig. 5.10 where the labels in the figure is given by the database.

Moreover, to verify that the proposal can indeed extract discriminative facial expression features for FER, the proposed method is compared with the existing new methods as shown in Table 5.2, which mainly focus on extracting discriminative features by various approaches which also is one of the motivation of the proposal. Concretely, the LDTP [7] is used to encodes the information of emotion-related features via ternary pattern and direction information. The exemplar-based SVM (ESVM) [18] is proposed to extract the informative features from expression data, and the histogram of oriented gradient is combined with the SVM to build the model. In [22], the salient facial patches (SFP) is selected by using appearance features that

Fig. 5.9 The confusion matrix on benchmark databases

	Angry	Disgust	Fear	Happiness	Neutral	Sadness	Surprise
Angry	.96	0	0	.04	0	0	0
Disgust	.09	.91	0	0	0	0	0
Fear	0	0	.85	0	0	.11	.04
Happiness	0	0	.04	.96	0	0	0
Neutral	0	0	0	0	1	0	0
Sadness	0	0	0	0	0	.96	.04
Surprise	0	0	0	.05	0	0	.95

(a) JAFFE

	Angry	Disgust	Fear	Happiness	Neutral	Sadness	Surprise
Angry	.83	.02	0	.02	.04	.09	0
Disgust	.03	.91	.02	.02	.02	0	0
Fear	.03	0	.80	.08	.03	.06	0
Happiness	0	.01	.03	.96	0	0	0
Neutral	.02	0	.04	.01	.89	.03	.01
Sadness	.05	0	.02	0	.03	.90	0
Surprise	0	0	0	0	.01	0	.99

(b) CK+

	Angry	Disgust	Fear	Happiness	Neutral	Sadness	Surprise
Angry	.50	.03	.05	.12	.11	.14	.05
Disgust	.23	.22	.03	.18	.17	.07	.10
Fear	.10	.02	.30	.14	.15	.15	.14
Happiness	.12	.02	.02	.62	.11	.08	.03
Neutral	.12	0	.04	.07	.60	.11	.06
Sadness	.10	.02	.05	.12	.15	.51	.05
Surprise	.19	.03	.08	.05	.19	.08	.38

(c) SFEW 2.0

Fig. 5.10 Some samples that are ambiguous in CK+ database

Table 5.2 Comparisons with the state-of-art approaches on benchmark databases

Database	Method	Accuracy (%)
JAFFE	CNN [12]	86.74
	SFP [22]	91.80
	WMDNN [35]	92.21
	ESVM [18]	92.53
	LDTP [7]	93.20
	WACNN	94.01
CK+	SDNMF [25]	72.90
	DAR [26]	85.00
	FRBM+ [21]	88.00
	LZMHI [27]	88.30*
	WCRAFM [32]	89.84
	WACNN	91.00
SFEW 2.0	AUDN [31]	30.14
	E-Gabor [23]	35.40
	MDP [24]	35.60
	CNNVA [8]	40.00
	WACNN	49.02

* Six basic expression+Contempt

comprises distinguish features for classifying each pair of expressions. And the subclass discriminant nonnegative matrix factorization (SDNMF) [25] unites suitable clustering-based differentiate criteria in the decompose cost function of nonnegative matrix factorization to find differentiate projections which can improve class divisibility in the dimensionality reduced space of projection. The DAR [26] is proposed to search distinguishing and correlated features. Two multilayer perceptrons is applied in DAR. In [27], the local Zernike moment with motion history image (LZMHI) is combined to incorporate velocity information of features. The enhanced Gabor (E-Gabor) feature descriptor is developed for producing high recognition accuracy of expression with low dimensionality in [23]. A monogenic directional pattern (MDP) is also proposed for extracting discriminative features [24].

Besides, the proposal is also compared with the other DNN based method to validate that the HGA used in WACNN can improve the performance of DCNN. The CNN combined with specific image preprocessing steps is proposed in [12]. Four-layer restricted Boltzmann machine (FRBM+) is proposed in [21], where FRBM is employed to catch label level and factor level dependency at the same time to identify multi-action units, and the nodes of facial expression are added in the middle visible layer of FRBM. In [32], a 14-layered CNN is established with the proposed WCRAFM to realize FER. And the WMDNN [35] ensembles the famous VGG16 network with the 3-layered CNN. Consequently, according to Table 5.2, it can be seen that the proposed WACNN is competitive compared with the methods focused on discriminative features extraction by various algorithms and the other methods based on DNN with more complex architecture. To learn better expression features representation, in [31] action units inspired deep networks (AUDN) is proposed. An 11-layered CNN with visual attention (CNNVA) in [8] is proposed for FER. It follows that the proposal can learn more discriminative features than that of other DNN-based method with the help of HGA.

5.7 Summary

To extract discriminative expression representations, a weight-adapted convolution neural network is proposed to get facial expression information which is inspired by the mechanism of human expression perception. The proposal preprocesses facial expression images first, then applies principal component analysis to extract low-level feature, finally the weight-adapted convolution neural network is exploited to learn representative feature and recognize them, so that the facial expression information is obtained.

Although deep learning has greatly improved the performance of facial expression recognition than some traditional algorithms, the improvement is limited. Therefore, various methods are adopted to assist deep learning in improving performance, such as design more complicated structure of deep learning, improve the feature extraction method, or fuse multiple different types of expression feature, and so on. Different from these methods, we focus on the optimization algorithm of deep learning more. The gradient-based algorithms are easily fall into local optimum, which does not guarantee optimal performance of neural network. However, despite some scholars have introduced evolutionary algorithms into the optimization of neural networks, most of them are relatively shallow neural networks and do not involve optimization of large-scale parameters. In this case, to make good use of the neural network's potential performance as much as possible, a hybrid genetic algorithm is developed to adaptively optimize the large-scale weights of deep convolution neural network. To speed up the convergence of the proposal, the weights optimized by accelerated gradient descent will be taken as an initial chromosome of the hybrid genetic algorithm, it also can be used as a priori knowledge. Moreover, the idea of novelty search is introduced to solve the issue of deception problem in the hybrid genetic algorithm, in such a way that the search space of hybrid genetic algorithm can be expended, and the chromosome with more novelty and the highest fitness is found during evolution, as a result, the HGA can jump out of local optimum and optimize the large scale parameters, the global search capability is also enhanced. To this end, the weight-adapted convolution neural network is established, which can extract more discriminative facial expression features for recognizing facial expression.

Experiments on JAFFE, CK+, and SFEW 2.0 databases are carried out by using k-fold cross validation to evaluate the effectiveness of the proposed algorithm. And the experimental results are better than that of the accelerated gradient descent with deep convolution neural networks, which shows that the proposal can find more optimal weights of the deep convolution neural networks by using the hybrid genetic algorithm, due to that the hybrid genetic algorithm has better global search capabilities than that of the gradient-based algorithm. Meanwhile, the experimental results are competitive compared with the existing new approaches aim to extract discriminative and are the deep neural network based methods. Furthermore, preliminary application experiment conducted on the emotional social robot system show that the wheeled robots are able to recognize the basic emotions such as happy, surprise, and so on.

For future work, the evolutionary algorithm for optimizing the both structure and weights of deep neural network for facial expression recognition in more complex environment will be explored. Moreover, people usually express their emotions through various modal information such as facial expressions, voices, body postures, etc. Hence, it is necessary to combine facial expression with other modalities to achieve more accurate emotion recognition. Nowadays, besides the ability to recognize emotions, how to make robots have the ability to understand emotions is becoming more and more important in human-robot interaction. Therefore, the understanding of emotional intention also will be considered in the future.

References

1. J. Jang, H. Cho, J. Kim, J. Lee, S. Yang, Facial attribute recognition by recurrent learning with visual fixation. IEEE Trans. Cybern. https://doi.org/10.1109/TCYB.2017.2782661.
2. L.F. Chen, M. Wu, M.T. Zhou, Z.T. Liu, J.H. She, K. Hirota, Dynamic emotion understanding in human-robot interaction based on two-layer fuzzy SVR-TS model. IEEE Trans. Syst., Man, Cybern.: Syst. https://doi.org/10.1109/TSMC.2017.2756447.
3. S. Poria, E. Cambria, R. Bajpai, A. Hussain, A review of affective computing: from unimodal analysis to multimodal fusion. Inf. Fusion **37**, 98–125 (2017)
4. A. Halder, A. Konar, R. Mandal, A. Chakraborty, P. Bhowmik, N.R. Pal, A.K. Nagar, General and interval type-2 fuzzy face-space approach to emotion recognition. IEEE Trans. Syst., Man, Cybern.: Syst. **43**(3), 587–605 (2013)
5. L.F. Chen, M. Wu, M.T. Zhou, J.H. She, F.Y. Dong, K. Hirota, Information-driven multi-robot behavior adaptation to emotional intention in human-robot interaction. IEEE Trans. Cogn. Dev. Syst. **10**(3), 647–658 (2018)
6. K. Mistry, L. Zhang, S.C. Neoh, C.P. Lim, B. Fielding, A Micro-GA embedded PSO feature selection approach to intelligent facial emotion recognition. IEEE Trans. Cybern. **47**(6), 1496–1509 (2016)

7. B. Ryu, A.R. Rivera, J. Kim, O. Chae, Local directional ternary pattern for facial expression recognition. IEEE Trans. Image Process. **26**(12), 6006–6018 (2017)
8. W. Sun, H. Zhao, Z. Jin, A visual attention based ROI detection method for facial expression recognition. Neurocomputing **296**, 12–22 (2018)
9. W. Samek, A. Binder, G. Montavon, S. Lapuschkin, K.R. Muller, Evaluating the visualization of what a deep neural network has learned. IEEE Trans. Neural Netw. Learn. Syst. **28**(11), 2660–2672 (2017)
10. Y. Sun, G.G. Yen, Z. Yi, Evolving unsupervised deep neural networks for learning meaningful representations. IEEE Trans. Evol. Comput. https://doi.org/10.1109/TEVC.2018.2808689.
11. Y. Lecun, Y. Bengio, G. Hinton, Deep learning. Nature **521**(7553), 436–444 (2015)
12. A.T. Lopes, E.D. Aguiar, A.F.D. Souza, T.O. Santosa, Facial expression recognition with convolutional neural networks: coping with few data and the training sample order. Pattern Recogn. **61**, 610–628 (2017)
13. R. Adolphs, Neural systems for recognizing emotion. Curr. Opin. Neurobiol. **12**(2), 169–177 (2002)
14. S. Strasser, J. Sheppard, N. Fortier, R. Goodman, Factored evolutionary algorithms. IEEE Trans. Evol. Comput. **21**(2), 281–293 (2017)
15. K.O. Stanley, R. Miikkulainen, Evolving neural networks through augmenting topologies. Evol. Comput. **10**(2), 99–127 (2002)
16. K.O. Stanley, Compositional pattern producing networks: a novel abstraction of development. Genet. Program Evolvable Mach. **8**(2), 131–162 (2007)
17. J. Lehman, K.O. Stanley, Abandoning objectives: evolution through the search for novelty alone. Evol. Comput. **19**(2), 189–223 (2011)
18. N. Farajzadeh, M. Hashemzadeh, Exemplar-based facial expression recognition. Inf. Sci. https://doi.org/10.1016/j.ins.2018.05.057.
19. M.N. Islam, M. Seera, C.K. Loo, A robust incremental clustering-based facial feature tracking. Appl. Soft Comput. **53**, 34–44 (2017)
20. L. Zhong, Q. Liu, P. Yang, J. Huang, D.N. Metaxas, Learning multiscale active facial patches for expression analysis. IEEE Trans. Cybern. **45**(8), 1499–1510 (2015)
21. S.F. Wang, S. Wu, G.Z. Peng, Q. Ji, Capturing feature and label relations simultaneously for multiple facial action unit recognition. IEEE Trans. Affect. Comput. https://doi.org/10.1109/TAFFC.2017.2737540.
22. S.L. Happy, A. Routray, Automatic facial expression recognition using features of salient facial patches. IEEE Trans. Affect. Comput. **6**(1), 1–12 (2015)
23. A.S. Alphonse, D. Dharma, Enhanced Gabor (E-Gabor), hypersphere-based normalization and pearson general kernel-based discriminant analysis for dimension reduction and classification of facial emotions. Expert Syst. Appl. **90**, 127–145 (2017)
24. A.S. Alphonse, D. Dharma, A novel monogenic directional pattern (MDP) and pseudo-voigt kernel for facilitating the identification of facial emotions. J. Vis. Commun. Image Represent. **49**, 459–470 (2017)
25. S. Nikitidis, A. Tefas, I. Pitas, Projected gradients for subclass discriminant nonnegative subspace learning. IEEE Trans. Cybern. **44**(12), 2806–2819 (2014)
26. C.O. Sakar, O. Kursun, Discriminative feature extraction by a neural implementation of canonical correlation analysis. IEEE Trans. Neural Netw. Learn. Syst. **28**(1), 164–176 (2017)
27. X.J. Fan, T. Tjahjadi, A dynamic framework based on local Zernike moment and motion history image for facial expression recognition. Pattern Recogn. **64**, 399–406 (2017)
28. Z. Sun, Z.P. Hu, M. Wang, S.H. Zhao, Discriminative feature learning-based pixel difference representation for facial expression recognition. IET Comput. Vision **11**(8), 675–682 (2017)

29. Y. Yang, Q.M.J. Wu, Y. Wang, Autoencoder with invertible functions for dimension reduction and image reconstruction. IEEE Trans. Syst., Man, Cybern.: Syst. **48**(7), 1065–1079 (2018)
30. L.F. Chen, M.T. Zhou, W.J. Su, M. Wu, J.H. She, K. Hirota, Softmax regression based deep sparse autoencoder network for facial emotion recognition in human-robot interaction. Inf. Sci. **2018**(428), 49–61 (2018)
31. M.Y. Liu, S.X. Li, S. Shan, X.L. Chen, AU-inspired deep networks for facial expression feature learning. Neurocomputing **159**, 126–136 (2015)
32. B.F. Wu, C.H. Lin, Adaptive feature mapping for customizing deep learning based facial expression recognition model. IEEE Access **6**, 12451–12461 (2018)
33. S.Y. Xie, H.F. Hu, Facial expression recognition using hierarchical features with deep comprehensive multi-patches aggregation convolutional neural networks. IEEE Trans. Multimed. https://doi.org/10.1109/TMM.2018.2844085.
34. A. Majumder, L. Behera, V.K. Subramanian, Automatic facial expression recognition system using deep network-based data fusion. IEEE Trans. Cybern. **48**(1), 103–104 (2018)
35. B. Yang, J. Cao, R. Ni, Y.Y. Zhang, Facial expression recognition using weighted mixture deep neural network based on double-channel facial images. IEEE Access **6**, 4630–4640 (2017)
36. W.J. Su, L.F. Chen, M. Wu et al., Nesterov accelerated gradient descent-based convolution neural network with dropout for facial expression recognition, in *Proceedings of the 2017 Asian Control Conference* (Gold Coast, Australia, 2018), pp. 1063–1068
37. H. Badem, A. Basturk, A. Caliskan et al., A new efficient training strategy for deep neural networks by hybridization of artificial bee colony and limited-memory BFGS optimization algorithms. Neurocomputing **266**, 506–526 (2017)
38. Y. Nesterov, A method of solving a convex programming problem with convergence rate $O(1/k^2)$. Sov. Math. Dokl. **27**(2), 372–376 (1983)
39. C.T. Lin, M. Prasad, A. Saxena, An improved polynomial neural network classifier using real-coded genetic algorithm. IEEE Trans. Syst., Man, Cybern.: Syst. **45**(11), 1389–1401 (2015)
40. J.T. Tsai, J.H. Chou, T.K. Liu, Tuning the structure and parameters of a neural network by using hybrid Taguchi-genetic algorithm. IEEE Trans. Neural Netw. **17**(1), 69–80 (2006)
41. E.P. Ijjina, K.M. Chalavadi, Human action recognition using genetic algorithms and convolutional neural networks. Pattern Recogn. **59**, 199–212 (2016)
42. C.F. Juang, Y.T. Yeh, Multiobjective evolution of biped robot gaits using advanced vontinuous ant-colony optimized recurrent neural networks. IEEE Trans. Cybern. **48**(6), 1910–1922 (2018)
43. M.G. Gong, J. Liu, H. Li, C. Qing, L.Z. Su, A multiobjective sparse feature learning model for deep neural networks. IEEE Trans. Neural Netw. Learn. Syst. **26**(12), 3263–3277 (2015)
44. J. Liu, M.G. Gong, Q.G. Miao, X.G. Wang, H. Li, Structure learning for deep neural networks based on multiobjective optimization. IEEE Trans. Neural Netw. Learn. Syst. **29**(6), 2450–2463 (2018)
45. N. Srivastava, G. Hinton, A. Krizhevsky, I. Sutskever, R. Salakhutdinov, Dropout: a simple way to prevent neural networks from overfitting. J. Mach. Learn. Res. **15**(1), 1929–1958 (2014)
46. P. Viola, M.J. Jones, Robust real-time face detection. Int. J. Comput. Vision **57**(2), 137–154 (2004)
47. J.H. Holland, *Adaptation in Natural and Artificial Systems* (MIT Press, Cambridge, 1992)
48. P. Ekman, W.V. Friesen, Constants across cultures in the face and emotion. J. Pers. Soc. Psychol. **17**(2), 124–129 (1971)
49. Cohn-Kanade (CK and CK+) database download site (2000) http://www.consortium.ri.cmu.edu/data/ck/
50. The Japanese female facial expression (JAFFE) Database (1998) http://www.kasrl.org/jaffe.html
51. Static facial expressions in the wild 2.0 (SFEW 2.0) database (2011) https://computervisiononline.com/dataset/1105138659

52. S. Murata, Y. Yamashita, H. Arie, T. Ogata, S. Sugano, J. Tani, Learning to perceive the world as probabilistic or deterministic via interaction with others: a neuro-robotics experiment. IEEE Trans. Neural Netw. Learn. Syst. **28**(4), 830–848 (2015)
53. J. Quintas, G.S. Martins, L. Santos, P. Menezes, J. Dias, Toward a context-aware human-robot interaction framework based on cognitive development. IEEE Trans. Syst., Man, Cybern.: Syst. https://doi.org/10.1109/TSMC.2018.2833384.
54. L.F. Chen, M.T. Zhou, M. Wu, J.H. She, Z.T. Liu, F.Y. Dong, K. Hirota, Three-layer weighted fuzzy support vector regression for emotional intention understanding in human-robot interaction. IEEE Trans. Fuzzy Syst. **26**(5), 2524–2538 (2018)

Chapter 6
Two-Layer Fuzzy Multiple Random Forest for Speech Emotion Recognition

The two-layer fuzzy multiple random forest (TLFMRF) is proposed for speech emotion recognition. When recognizing speech emotion, there are some problems. One is that feature extraction relies on personalized features. The other is that emotion recognition doesn't consider the differences among different categories of people. In the proposal, personalized and non-personalized features are fused for speech emotion recognition. High dimensional emotional features are divided into different subclasses by adopting the fuzzy C-means clustering algorithm, and multiple random forest is used to recognize different emotional states. Finally, a TLFMRF is established. Moreover, a separate classification of certain emotions which are difficult to recognize to some extent is conducted. The results show that the TLFMRF can identify emotions in a stable manner.

6.1 Introduction

Robotics has been widely developed recently, and people hope that human-robot interaction (HRI) becomes more humanized and naturalized, even robot understands human intention [1, 2]. Moreover, the robot is expected to have the ability to express emotions and exhibit some reactions [3]. Affective computing is attracting more and more attention. Emotions are an important bond in human-robot interaction, and they can be perceived by speech signals [4], facial expressions [5], physiological signal [6] such as electrocardiograph (ECG), blood pressure and electroencephalogram (EEG), and body posture, among others. Emotion recognition has received much attention in recent years [7–9]. Speech signals as a main way of affective computing in HRI, have been applied in HRI [10–12]. Speech emotion recognition (SER), which is defined as extracting the emotional state of a speaker from his or her speech, is used to acquire useful semantic information from speech, furthermore, improves the performance of HRI systems.

L. Chen et al., *Emotion Recognition and Understanding for Emotional Human-Robot Interaction Systems*, Studies in Computational Intelligence 926,
https://doi.org/10.1007/978-3-030-61577-2_6

SER system consists of two major stages. One is the choice of suitable features for speech representation, the other is a design of an appropriate classification [13]. Feature extraction focuses on emotion-relevant features from voice, and a feature classification consists of a training and a test phase for identifying the type of emotion. Over the past decades, various features have been investigated and applied for SER [14, 15], and some optimization algorithms are also been applied in features selection, such as particle swarm optimization (PSO) [16], ant colony optimization (ACO) [17], and so on [18].

Especially, the characteristics of speaker and language can improve SER performance. According to language characteristics, speech features can be divided into prosodic features, sound quality features, and spectral features. However, many studies considered acoustic or/and prosodic features, such as pitch, intensity, voice quality features, spectrum and cepstrum [19]. Among these studies, the global statistics over the low level descriptors (LLDs), e.g., fundamental frequency (F_0), durations, intensities, Mel frequency cepstrals (MFCCs) [13], had achieved dominant superiority. Wu et al. [20] used acoustic-prosodic information and semantic labels for SER. Eyben et al. [21] used the openSMILE toolkit to achieve the extraction of short-term acoustic features, such as pitch, energy, F_0, time duration and MFCC. In this chapter, F_0, root-mean-square signal frame energy (RMS energy), cross-zero rate (zcr), harmonic noise ratio, MFCC that are proposed for SER. According to speaker, some studies just focused on personalized speech features [22], in general, emotional speech data expressed by different speakers demonstrated large variations in acoustic characteristics, even if they intend to express the same emotion. Several pairs of representative emotions tend to have similar acoustic characteristics. For example, voices of sadness and boredom have similar characteristics, indicating a large overlap among acoustic features [23]. Thus, speaker-independent speech, namely, non-personalized speech emotional features, which do not rely on the speaker, are adopted for SER [24]. As a result, personalized and non-personalized speech emotional features are proposed for SER in this chapter. This is a strong motivation of the research.

Furthermore, a suitable classifier for identification of emotional states is designed. Representative classifiers have been used for the task of SER, including hidden Markov model (HMM) [25], the Gaussian mixture model (GMM) [26], support vector machine (SVM) [27, 28], and artificial neural network (ANN) [29, 30]. In the stage of speech feature recognition, many researchers analysis the characteristics and distribution of speech emotion features, and hierarchical classifier is used to improve the performance of classifier. According to the hierarchical classification of prosodic information and semantic tags, Wu et al. [20] obtained the final result by weighted integration into the semantic tag, and tested under natural sound environment. Yuncu et al. [31] produced the easily distinguishable degree of the emotion categories by choosing the feature set, the size of the confusion between different emotions. The corresponding two decision tree were constructed, and the emotional states recognition were completed layer by layer. For considering the lack of feature sets, Sheikhan et al. [32] presented a fuzzy neural-SVM recognition method. The identification results based on different feature sets are divided into three layers to make fuzzy decision step by step, then the final recognition results are obtained. At present, the combination of standard methods, such as fusion, ensemble or hierarchical classifiers, has become the key point in SER.

Enrique et al. [33] developed a novel ensemble classifier that consists of multiple standard classifiers where the SVM is used to deal with multiple languages, and it was tested on never-seen languages. Sarker et al. [34] adopted neural network, decision tree, SVM and k-nearest neighbor (KNN) to classify the test data. Due to the lack of specific studies on high dimensional feature data. Breiman et al. [35] proposed the random forest (RF), which is an algorithm based on classification trees, and a large number of independent variables which may be up to several thousand could be handled. In addition, RF is used as a machine learning algorithm, both for individual feature sets and for decision-level fusion [36, 37]. Recently, RF have been used for natural language recognition [38, 39]. Kondo et al. [38] suggested that when compared with ANN, support vector regression (SVR), logistic regression (LR), RF had better performance.

In this chapter, RF is adopted to recognize emotional speech, including emotions of angry, fear, happy, neutral, sad and surprise. Personalized features and non-personalized features are fused for SER. Identification information of human (i.e., gender, province, and age) have important influence on emotional intention understanding. Therefore, it can be seen that identification information has a certain influence on SER, and this is the other motivation of the proposal. According to the identification information, such as gender and age, the feature data can be divided into different subclasses by adopting the Fuzzy C-means (FCM) clustering, which the Euclidean distance and membership functions are used to cluster the speech data based on the characteristics of it. Then, two-layer fuzzy multiple random forest (TLFMRF) is proposed, where the decision-tree and Bootstrap method are employed in the RF to recognize the speech features. Finally, our model is established, and the confusion matrix is formed. Moreover, a separate classification of some certain emotions that are difficult recognition to some extent is carried out in each of the multiple classifications.

In Sect. 6.2, TLFMRF for SER is introduced. The experimental simulations and analysis including a further application are presented in Sect. 6.3. In Sect. 6.4, a summary and prospective of the proposal are covered.

6.2 Feature Extraction

The feature sets are computed using openSMILE [21] toolkit (version 2.3). 16 basic features and their 1st derivatives are extracted as fundamental features. Derivative features are less affected by different person, which are seen as non-personalized features. 12 statistics values of these fundamental features are calculated. Therefore, about 384 features are extracted in total, as shown in Table 6.1. According to this method, the personalized features and non-personalized features are obtained.

In the speech emotional features, the zcr (zero-crossing rate) refers to the fact that the signal pattern we are talking through passes through the zero-level record. The zcr of the speech signal $x(m)$ is computed as

Table 6.1 Emotional speech features

Index	16 Basic features	12 statistic values
Personalized features	F_0, zcr, RMS energy, Harmonic noise ratio, MFCC 1-12	max, min, average, std, range, maxpos, minpos, linregc1, linregc2, linregerrQ, kurtosis, skewness
Non-personalized features	1st order delta coefficient of 16 basic features	

$$Z = \frac{1}{2}\sum_{m=0}^{N-1} |\text{sgn}[x(m)] - \text{sgn}[x(m-1)]| \tag{6.1}$$

where $sgn\,[\cdot]$ is defined as

$$\text{sgn}[x] = \begin{cases} 1, & (x \geq 0) \\ -1, & (x < 0) \end{cases} \tag{6.2}$$

And MFCC can be calculated by the following steps:

$Step1$: Improve the speaker's voice information by the Hamming window and the framing. After such pre-processing, the FFT is implemented to obtain the spectrum.

$Step2$: Square the result of $Step1$, then pass it through the corresponding triangle filter, and then evenly arrange the center frequency according to the Mel frequency scale. The center frequency of the bandpass filter is divided by the interval of 150Mel and the bandwidth of 300 Mel. Suppose the number of filters is M, and the output frequency after filtering is $X(k), k = 1, 2, \ldots, M$.

$Step3$: Calculate the logarithm of the output of the bandpass filter in $Step2$, and then calculate the obtained power log spectrum by the following formula to obtain K MFCC ($K = 12 - 16$). According to the symmetry, the transformation can be simplified as follows:

$$C_n = \sum_{k=1}^{n} \log Y(k) \cos[\pi (k - 0.5) n/N], n = 1, 2, \ldots, N \tag{6.3}$$

where N represents the number of filters, and C_n is the filtered output.

The personalized features exhibit a positive impact on the SER of a certain person, but in the case of unfamiliar speaker without database, the emotion recognition rate is not very high. Derivative-based non-personalized features can solve this problem. Therefore, both personalized features and non-personalized features are used for SER in this chapter. Then, by using FCM, the training set are clustered into multi subclasses. Finally, RF is used to identify the emotion of the selected speech features. Six basic emotion are employed here, i.e., surprise, happy, sad, angry, fear, and neutral [40]. The framework of TLFMRF for SER is shown in Fig. 6.1.

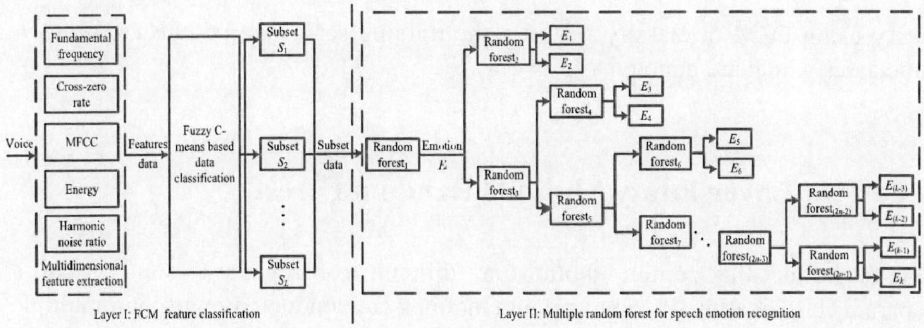

Fig. 6.1 The framework of two-layer fuzzy multiple random forest for speech emotion recognition

6.3 Fuzzy-c-Means Based Features Classification

The Fuzzy-c-means(FCM) is used for data clustering algorithm. Based on the method of bootstrap, the training sample data set D is obtained, i.e., $D = [\, y_1\ y_2\ \cdots\ y_N\,]^T$ and $y_o = [\, y_{o1}\ y_{o2}\ \cdots\ y_{o384}\,]$, $o = 1, 2, \ldots, N$, where N is the number of sample data. The FCM method is an iterative clustering algorithm to partition l samples into c clusters by minimizing the following objective function,

$$\begin{aligned} &\min J_m(U, V) = \sum_{k=1}^{L} \sum_{o=1}^{N} (\mu_{io})^m D_{ko}^2 \\ &D_{ko}^2 = \|y_o - c_k\|^2 \\ &\text{s.t.} \begin{cases} \sum_{k=1}^{L} \mu_{ko} = 1 \\ 0 < \mu_{ko} < 1 \end{cases} k = 1, \ldots, L,\ o = 1, \ldots, N \end{aligned} \tag{6.4}$$

where μ_{ko} is the membership value of the oth sample in the kth cluster, U is the related fuzzy partition matrix consisting of μ_{ko}, $V = (c_1, c_2, \ldots, c_L)$ is the cluster center matrix, L is the number of clusters, m is the fuzzification exponent and usually $m = 2$, D_{ko} is the Euclidean distance between oth sample y_o and kth cluster center c_k.

To minimize J_m, the following update equations are used

$$\mu_{ko} = \frac{1}{\sum_{f=1}^{L} (D_{ko}/D_{fo})^{2/(m-1)}} \tag{6.5}$$

$$c_k = \frac{\sum_{o=1}^{N} (\mu_{ko})^m y_o}{\sum_{o=1}^{N} (\mu_{ko})^m} \tag{6.6}$$

By using FCM to classify features, the training set S can be clustered into L subclasses which are denoted as $S = \{S_1, S_2, \ldots, S_L\}$.

6.4 Two-Layer Fuzzy Multiple Random Forest

Due to the fact that certain emotions are difficult to recognize to some extent, a separate classification of some certain emotions are conducted by using a multiple random forest algorithm. It includes K emotional states, and takes out 2 categories of emotions that are relatively difficult to recognize each time. It can be concluded that M random forest is needed, which is given by

$$M_{RF} = 2\left|\frac{K}{2}\right| - 1, K = 1, 2, \ldots n \tag{6.7}$$

where M is the number of random forests, and K is the number of emotional states. In this chapter, emotional corpus database includes 6 basic emotions, and the proposed method solves the mutual interference between emotions so that the recognition rate of different emotion has been greatly improved. TLFMRF algorithm includes four steps.

Step1: Extract feature data through speech signal pre-processing by using openSMILE.

Step2: By using FCM, taking into account the impact of identification information on emotions, the training set is clustered into L subclasses.

Step3: Train classifier of RF. As a result, a total of 5 classifiers are trained. Classifier 1 distinguishes between sad, fear and others. Classifier 2 distinguishes between sad and fear. Classifier 3 distinguishes others. Classifier 4 distinguishes emotions of happy and neutral. Classifier 5 distinguishes angry and surprise. The structure of multiple random forest for SER is shown as Fig. 6.2.

Step4: Use trained TLFMRF for classification of six basic emotion, and integrate L subclasses into the classification result.

6.5 Experiments

In this part, the data settings and experimental environment settings are explained. The simulation experiment is carried out and the experimental results are analyzed.

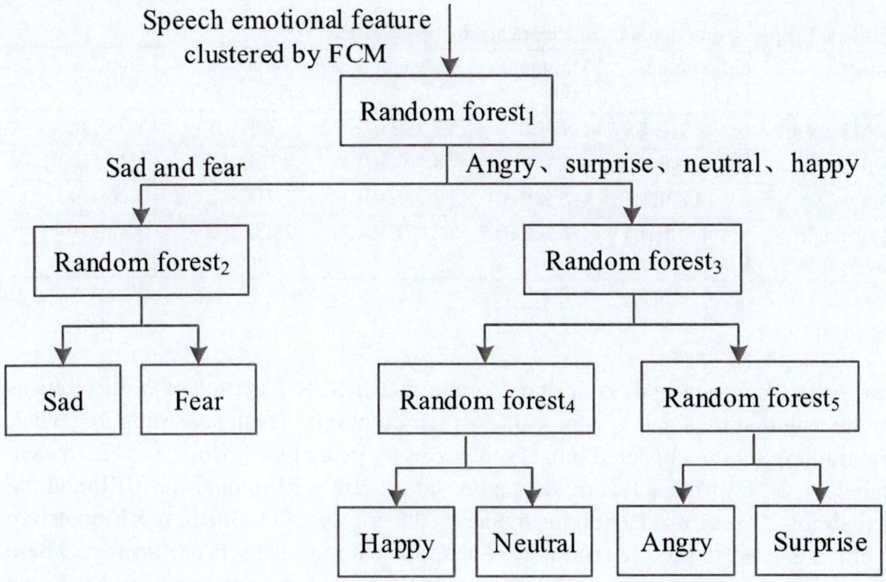

Fig. 6.2 The structure of multiple random forest for speech emotion recognition

6.5.1 Data Setting

CASIA corpus [40] is used for the experiment in which emotion speeches of four people (2 male, 2 female) are recorded. They speak the same 300 basic emotional short utterances using six basic emotion, i.e., angry, fear, happy, neutral, sad, and surprise that includes a total of 7200 segments of speech. Speech emotion features set are extracted by using openSMILE toolkit [21]. Some basic features such as RMS energy, zcr, harmonic noise ratio, MFCC are obtained as shown in Table 6.2. In order to realize SER that does not depend on speaker and natural environment, speech emotional feature is divided into personalized features and non-personalized features. In the personalized features aspect, MFCC include 12 spectral energy dynamic coefficients on equal frequency bands. These basic features are calculated by 12 statistic values. Being different from personalized features, the non-personalized speech emotional feature is used to eliminate the influence of different speakers' individual by introducing rate of change.

6.5.2 Environment Setting

After completing the preprocessing of speech emotional feature sets, 3600 * 384 dimensional eigenvectors are obtained from one man and one women, where each eigenvector corresponds to a label (1-angry, 2-fear, 3-happy, 4-neutral, 5-sad, 6-

Table 6.2 Speech emotional feature extracted by openSMILE

Index	Maximum value	Minimum value	Mean	Maximum value	Slope
RMS energy	7.51E-05	1.61E-01	2.25E+02	4.30E+01	1.98E-02
MFCC	7.38E+00	−3.62E+01	4.36E+01	3.73E+02	−1.15E+01
zcr	4.82E+0	−2.58E+01	2.23E+01	3.71E+02	2.26E+00
Harmonic noise ratio	4.30E+01	−2.24E+01	2.07E+0	2.43E+02	9.43E-01

surprise). In fact, age and gender have some effect on SER as a result of the differences in the way that men and women express their emotions. Then, according to gender, the training set are clustered into L subclasses by using FCM, while $L = 2$. In each subclass, 80% of these feature data are used to train a RF model for SER, and the remaining 20% are used to test the model. In the process of establishing RF, bootstrap method is used to sample randomly. Then, 500 samples subsets are formed, where the same sample may be selected repeatedly. With that, a decision tree is generated by training on each subset of samples, and a RF model is formed. As a result, RF algorithm is used for classification of six basic emotion. Due to a low recognition rate of certain emotions, multiple random forest (MRF) algorithm is adopted to identify those certain emotions that are difficult to distinguish.

6.5.3 *Simulations and Analysis*

To verify the effectiveness of SER, different classification model, which are back propagation neural network (BPNN), RF, and TLFMRF, are used. To verify the validity of the model, the five-fold cross validation method is used to verify the algorithm, namely, the data set is evenly divided into 5 parts. Then, the experiments are carried out in different ways. In the end, the results of 5 cycles are output, where the cyclic variable is k.

The comparison of the three methods of SER results are shown in Table 6.3. In the first case, the BPNN algorithm is adopted, which includes 3 layer neural network, the number of hidden layer nodes is set to 100, the activation function is sigmoid function, and the number of nodes in the output layer is 6. According to the BPNN algorithm, the confusion matrix of SER result by using BPNN is shown in Fig. 6.3, and the above results are obtained by cross validation. From the above results, the average recognition rate is 81.75%.

In the second case, the RF algorithm is adopted firstly. The confusion matrix of SER result by using RF is shown in Fig. 6.4. Similarly, the above results are obtained by cross validation, the average recognition rate is 79.08%. It is seen from the above results that the average recognition rate of RF is low compared with BPNN. Thus, TLFMRF is proposed, while $L = 2$. Results show that SER average result obtains

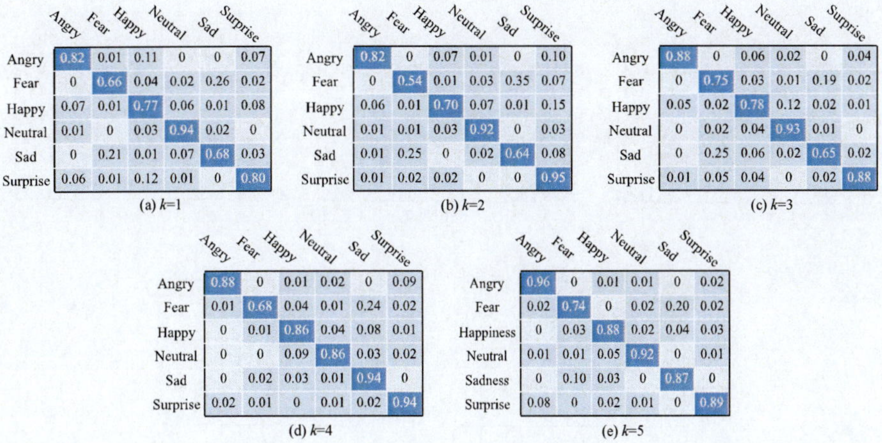

Fig. 6.3 Confusion matrix of recognition results by using BPNN

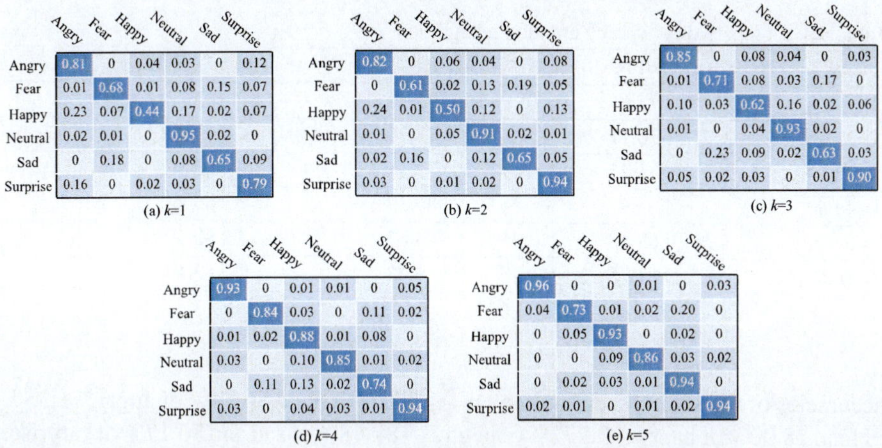

Fig. 6.4 Confusion matrix of recognition results by using random forest

accuracy of 83.14% by using TLFMRF, which is 4.06% higher than that of RF. Meantime, compared with the BPNN, the proposed TLFMRF is obviously high up to 1.39%, and the confusion matrix of SER results by using TLFMRF is shown in Fig. 6.5. Similarly, the above results are obtained by cross validation. According to the comparison of 3 methods of SER results, it is obvious that the proposed method is better and the model is relatively more stable that is of great importance in HRI.

Moreover, according to Figs. 6.3, 6.4, and 6.5, the average accuracies of the six basic emotion by using BPNN are 86.83% of angry, 67.49% of fear, 79.83% of happy, 93.17% of neutral, 75.50% of sad and 89.33% of surprise; the average accuracies of the six basic emotion by using RF are 87.34% of angry, 71.33% of fear, 67.19% of happy, 89.00% of neutral, 74.50% of sad and 89.50% of surprise; the average

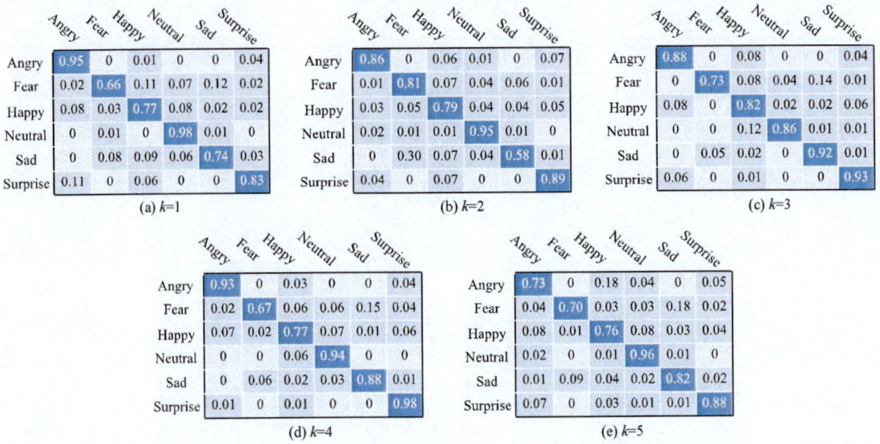

Fig. 6.5 Confusion matrix of recognition results by using two-layer fuzzy multiple random forest

Table 6.3 Comparison of speech emotion recognition

Index	BPNN	RF	TLFMRF
k=1 (%)	77.64	71.94	82.22
k=2 (%)	76.11	73.75	81.39
k=3 (%)	81.25	77.36	85.56
k=4 (%)	85.97	86.39	85.83
k=5 (%)	87.78	85.87	80.69
Average (%)	81.75	79.08	83.14

accuracies of the six basic emotion by using TLFMRF are 86.84% of angry, 71.33% of fear, 78.00% of happy, 93.772% of neutral, 78.648% of sad and 80.17% of surprise. Therefore, it can be seen that the accuracies of every emotion by using the proposed method is roughly higher than that of BPNN and RF. The accuracies of fear, happy and sad are relatively lower in the six basic emotions by using the above 3 methods. Between fear and sad, there is also a relatively high confusions by using the 3 methods. This is due to that the fear and sad are similar when expressed by speech signals.

According to the SER results of 3 methods, it is seen that the average recognition rate of TLFMRF in SER is higher than that of BPNN and RF. Moreover, TLFMRF have great advantages in dealing with high-dimensional data, where identification information is embodied. As a result, age and gender are taken into account for data classification. In addition, the computation time for getting the results in Table 6.3, in which the computation time of TLFMRF, RF, and BPNN are 0.0707s, 0.0196s, and 0.0024s, respectively. Although the proposed algorithm has the longest computation time, the computation time is still at sub-second range.

6.6 Summary

A TLFMRF is developed to recognize emotional states by involving speech signals. It mainly solves the problems to include the choice of features and a classification method identification. In the aspect of speech emotional feature extraction, we adopt the non-personalization speech emotional feature based on derivative to supplement the traditional speech personalized emotional characteristics, and realize the universal and negotiability emotional characteristics. With regard to speech emotional feature classification, TLFMRF is adopted to deal with high dimension correlation features and improve the recognition result. In TLFMRF, the FCM is adopted first to divide the feature data into different subclasses according to the identification information by using Euclidean distance and membership functions. Next, the multiple RF is employed by using decision-tree and Bootstrap method to recognize these feature data.

The proposed TLFMRF considers sufficiently the impact of futures, the novelty are not only it that extracts the non-personalized features by taking the derivative of personalized features, but also it divides the high dimension feature data into different subset data in such a way that the computational dimension is reduced and characteristics of each subset data are similar to ensure the learning efficiency. The main features are summarized as follows:

(1) To avoid the problem that feature extraction relies on the personalized features, personalized features and non-personalized features are extracted and fused.

(2) By considering that emotion recognition does not take into account different categories of people, multiple random forest is adopted to recognize different emotional states.

(3) Two-layer fuzzy multiple random forest is proposed to improve recognition rate. Since the high dimensional correlation features are divided into different subclasses by adopting the fuzzy C-means, separate classification of emotions is carried out in each of the random forest, in such a way that indistinguishable emotions are identified and the recognition rates are improved.

In order to verify the validity of the TLFMRF, experiments on CASIA corpus is carried out by using k-fold cross validation, and experimental results show the recognition accuracy of the proposal with 83.14% are higher than that of the baseline algorithms, such as BPNN with the accuracy of 81.75%, RF with the accuracy of 79.08%.

For further research, intelligent optimization algorithms can be employed in the TLFMRF to further improve the performance of recognition [41, 42].

References

1. L.F. Chen, M. Wu, M.T. Zhou, Z.T. Liu, J.H. She, K. Hirota, Dynamic emotion understanding in human-robot interaction based on two-layer fuzzy SVR-TS model. IEEE Trans. Syst. Man Cybern.: Syst. **50**(2), 490–501 (2020)
2. L.F. Chen, Z.T. Liu, M. Wu, M. Ding, F.Y. Dong, K. Hirota, Emotion-age-gender-nationality based intention understanding in human-robot interaction using two-layer fuzzy support vector regression. Int. J. Soc. Robot. **7**(5), 709–729 (2015)
3. L.F. Chen, M. Wu, M.T. Zhou, J.H. She, F.Y. Dong, K. Hirota, Information-driven multi-robot behavior adaptation to emotional intention in human-robot interaction. IEEE Trans. Cognit. Developmen. Syst. **10**(3), 647–658 (2018)
4. L. Devillers, M. Tahon, M.A. Sehili et al., Inference of human beings' emotional states from speech in human-robot interactions. Int. J. Soc. Robot. **7**(4), 451–463 (2015)
5. L.F. Chen, M.T. Zhou, W.J. Su, M. Wu, J.H. She, K. Hirota, Softmax regression based deep sparse autoencoder network for facial emotion recognition in human-robot interaction. Inf. Sci. **428**, 49–61 (2018)
6. J. Kim, E. André, Emotion recognition based on physiological changes in music listening. IEEE Trans. Pattern Anal. Mach. Intell. **30**(12), 2067–2083 (2008)
7. F.Y. Leu, J.C. Liu, Y.T. Hsu et al., The simulation of an emotional robot implemented with fuzzy logic. Soft Comput. **18**(9), 1729–1743 (2014)
8. E.M. Albornoz, D.H. Milone, H.L. Rufiner, Feature extraction based on bio-inspired model for robust emotion recognition. Soft Comput. **21**(17), 5145–5158 (2017)
9. V.P. Gonçalves, G.T. Giancristofaro, G.P.R. Filho et al., Assessing users' emotion at interaction time: a multimodal approach with multiple sensors. Soft Comput. **21**(18), 5309–5323 (2017)
10. M.T. Zhou, L.F. Chen, J.P. Xu, X.H. Cheng, M. Wu, W.H. Cao, J.H. She, K. Hirota, FCM-based multiple random forest for speech emotion recognition, in *Proceedings of the 5th International Workshop on Advanced Computational Intelligence and Intelligent Informatics*, 1-2-4-1-6 (2017)
11. S. Zhang, X. Zhao, B. Lei, Speech emotion recognition using an enhanced kernel isomap for human-robot interaction. Int. J. Adv. Robot. Syst. **10**(2), 1–7 (2013)
12. B.W. Schuller, A.M. Batliner, *Computational Paralinguistics: Emotion, Affect and Personality in Speech and Language Processing* (John Wiley & Sons Inc, New York)
13. M.E. Ayadi, M.S. Kamel, F. Karray, Survey on speech emotion recognition: features, classification schemes, and databases. Pattern Recogn. **44**(3), 572–587 (2011)
14. P. Song, S.F. Ou, Z.B. Du et al., Learning corpus-invariant discriminant feature representations for speech emotion recognition. IEICE Trans. Inf. & Syst. **E100-D** (5), 1136–1139 (2017)
15. K. Hakhyun, E. Hokim, Y. Keunkwak, Emotional feature extraction method based on the concentration of phoneme influence for human-robot interaction. Adv. Robot. **24**(1–2), 47–67 (2010)
16. W. Deng, R. Yao, H. Zhao et al., A novel intelligent diagnosis method using optimal LS-SVM with improved PSO algorithm. Soft Comput. **2–4**, 1–18 (2017)
17. W. Deng, H.M. Zhao, L. Zou et al., A novel collaborative optimization algorithm in solving complex optimization problems. Soft Comput. **21**(15), 4387–4398 (2017)
18. W. Deng, S. Zhang, H. Zhao et al., A novel fault diagnosis method based on integrating empirical wavelet transform and fuzzy entropy for motor bearing. IEEE Access **6**(1), 35042–35056 (2018)
19. B. Schuller, S. Steidl, A. Batliner, The INTERSPEECH emotion challenge, in *Proceedings of INTERSPEECH*, pp. 312–315 (2009)
20. C.H. Wu, W.B. Liang, Emotion recognition of affective speech based on multiple classifiers using acoustic-prosodic information and semantic labels. IEEE Trans. Affect. Comput. **2**(1), 10–21 (2010)
21. F. Eyben, M. Wöllmer, A. Graves et al., Online emotion recognition in a 3-D activation-valence-time continuum using acoustic and linguistic cues. J. Multimodal User Interfaces **3**(1–2), 7–19 (2010)

22. J.B. Kim, J.S. Park, Multistage data selection-based unsupervised speaker adaptation for personalized speech emotion recognition. Eng. Appl. Artif. Intell. **52**(C), 126–134 (2016)
23. J.S. Park, J.H. Kim, Y.H. Oh, Feature vector classification based speech emotion recognition for service robots. IEEE Trans. Cons. Electron. **55**(3), 1590–1596 (2009)
24. E.H. Kim, K.H. Hyun, S.H. Kim et al., Improved emotion recognition with a novel speaker-independent feature. IEEE/ASME Trans. Mechatron. **14**(3), 317–325 (2009)
25. M. Deriche, A.H.A. Absa, A two-stage hierarchical bilingual emotion recognition system using a hidden Markov model and neural networks. Arabian J. Sci. & Eng. **42**(12), 5231–5249 (2017)
26. A. Mohamed, G.E. Dahl, G. Hinton, Acoustic modeling using deep belief networks. IEEE Trans. Audio Speech & Lang. Process. **20**(1), 14–22 (2012)
27. A.D. Dileep, C.C. Sekhar, GMM-based intermediate matching kernel for classification of varying length patterns of long duration speech using support vector machines. IEEE Trans. Neural Netw. & Learn. Syst. **25**(8), 1421–1432 (2014)
28. L.F. Chen, M.T. Zhou, M. Wu, J.H. She, Z.T. Liu, F.Y. Dong, K. Hirota, Three-layer weighted fuzzy support vector regression for emotional intention understanding in human-robot interaction. IEEE Trans. Fuzzy Syst. **26**(5), 2524–2538 (2018)
29. J. Deng, Z. Zhang, E. Marchi et al., *Sparse autoencoder-based feature transfer learning for speech emotion recognition, in Proceedings of Humaine Association Conference on Affective Computing and Intelligent Interaction* (Geneva, Switzerland, 2013), pp. 511–516
30. H.M. Fayek, M. Lech, L. Cavedon, Evaluating deep learning architectures for speech emotion recognition. Neural Netw. **92**, 62–68 (2017)
31. E. Yuncu, H. Hacihabiboglu, C. Bozsahin, Automatic speech emotion recognition using auditory models with binary decision tree and SVM, in *Proceedings of International Conference on Pattern Recognition*, pp. 773–778 (2014)
32. M. Sheikhan, M. Bejani, D. Gharavian, Modular neural-SVM scheme for speech emotion recognition using ANOVA feature selection method. Neural Comput. Appl. **23**(1), 215–227 (2013)
33. M.E. Albornoz, D. Milone, Emotion recognition in never-seen languages using a novel ensemble method with emotion profiles. IEEE Trans. Affect. Comput. **8**(99), 1–11 (2016)
34. Y. Sun, G. Wen, Ensemble softmax regression model for speech emotion recognition. Multimedia Tools & Appl. **76**(6), 8305–8328 (2016)
35. L. Breiman, Random forests. Mach. Learn. **45**, 5–32 (2010)
36. E. Vaiciukynas, A. Verikas, A. Gelzinis et al., Detecting Parkinson's disease from sustained phonation and speech signals. Plos One **12**(10), 1–16 (2017)
37. R. Genuer, J.M. Poggi, C. Tuleau-Malot, Variable selection using random forests. Elsevier Science Inc **31**(14), 2225–2236 (2010)
38. K. Kondo, K. Taira, K. Kondo et al., Estimation of binaural speech intelligibility using machine learning. Appl. Acoust. **129**, 408–416 (2018)
39. T. Iliou, C.N. Anagnostopoulos, *Comparison of different classifiers for emotion recognition, in Proceedings of Panhellenic Conference on Informatics* (Corfu, Greece, 2009), pp. 102–106
40. CASIA Chinese Emotion Corpus. [Online], http://www.chineseldc.org/resourceinfo.php?rid=76. Accessed 11 June 2008
41. W. Deng, R. Chen, B. He et al., A novel two-stage hybrid swarm intelligence optimization algorithm and application. Soft Comput. **16**(10), 1707–1722 (2012)
42. W. Deng, H. Zhao, X. Yang et al., Study on an improved adaptive PSO algorithm for solving multi-objective gate assignment. Appl. Soft Comput. **59**, 288–302 (2017)

Chapter 7
Two-Stage Fuzzy Fusion Based-Convolution Neural Network for Dynamic Emotion Recognition

The two-stage fuzzy fusion based-convolution neural network is proposed for dynamic emotion recognition by using both facial expression and speech modalities, which not only can extract discriminative emotion features which contain spatio-temporal information, but can also effectively fuse facial expression and speech modalities. Moreover, the proposal is able to handle situations where the contributions of each modality data to emotion recognition are very imbalanced. The local binary patterns coming from three orthogonal planes and spectrogram are considered first to extract low-level dynamic emotion, so that the spatio-temporal information of these modalities can be obtained. To reveal more discriminative features, two deep convolution neural networks are constructed to extract high-level emotion semantic features. Moreover, the two stage fuzzy fusion strategy is developed by integrating canonical correlation analysis and fuzzy broad learning system, so as to take into account the correlation and difference between different modal features, as well as handle the ambiguity of emotional state information.

7.1 Introduction

Emotion recognition is significant for social communication in our daily life, and emotions play crucial roles in determining human behavior [1]. Consequently, realizing emotion recognition is important to achieve the harmonious and natural Human-Robot Interaction (HRI). Nonverbal communication, such as facial expressions and voice tone, can emphasize the implications of conversation [2]. Therefore, it is beneficial to realize emotion recognition based on nonverbal communication. In addition, both facial expression and speech exhibit complementary effects in expressing emotions. To this end, the modalities of facial expression and speech are considered to achieve dynamic emotion recognition. However, recognizing human emotions with multimodal information remains a challenging task for extracting and fusing repre-

L. Chen et al., *Emotion Recognition and Understanding for Emotional Human-Robot Interaction Systems*, Studies in Computational Intelligence 926,
https://doi.org/10.1007/978-3-030-61577-2_7

sentative facial expression and speech features which are discriminative of imperceptible changes in emotion [3].

In emotion recognition with multimodal information, the emotional feature extraction and multimodal fusion are the most important steps [4]. As for feature extraction of facial expression, a number of studies were focused on static expression feature extraction [3, 5–7]. Nevertheless, compared with static expression feature extraction, dynamic expression feature extraction can contain minute information about the change of facial expression which contains spatio-temporal information. The Local Binary Patterns from Three Orthogonal Planes (LBP-TOP) [8] are robust to image transformation (e.g., rotation), insensitive to illumination changes, and enjoy computational simplicity, so that the LBP-TOP is applied to extract low-level dynamic features of facial expression. As for feature extraction of speech, it mainly includes prosodic features, spectral features, and acoustic features [9]. In the field of speech signal processing, spectral correlation features are regarded as an expression of the correlation between vocal tract shape change and articulator movement. Howbeit, traditional spectral correlation features are confined to the frequency domain, ignoring some information in the time domain. Compared with traditional spectral correlation features, spectrogram can break through the singularity of traditional spectral related features due to the fact that it comprehensively considers the frequency and time information and is modeled as an image [10]. Moreover, more emotion-related information can be extracted from the image feature descriptor of spectrogram [11, 12]. However, the feature extraction algorithms mentioned above can be summarized as low-level feature extraction, which have been shown that they are not discriminative enough to recognize the emotions [4, 13]. Deep learning can form more abstract high-level representation of data by combining low layer features to discover the distributed feature representation of the data [14, 15]. Hence, the Deep Convolution Neural Network (DCNN) is introduced to extract high-level emotion semantic features.

The choice of the multimodal information fusion method has a significant influence on the effect and performance of multimodal emotion recognition. Fusion of multimodal data can improve the accuracy of overall results or decisions [16]. The main multimodal fusion methods are: feature-based fusion, decision-based fusion, and hybrid fusion [17, 18]. The feature-based fusion can produce feature information needed in decision analysis to the greatest extent. Different from feature-based fusion, decision-based fusion regards different modal of emotional features as independent, and considers the importance of different modal emotional feature data for emotion recognition. Hybrid fusion attempts to exploit the advantages of feature-based and decision-based fusion methods in a common framework. In order to take into account the correlation and difference between different modal features at the same time, a novel hierarchical fusion structure of Two-Stage Fuzzy Fusion Strategy (TSFFS) is proposed. At the first stage, the feature-based fusion realized by using Canonical Correlation Analysis (CCA) is conducted, where high-level semantic emotional features of facial expression and speech data are fused. It can effectively remove redundancy between features, find the internal relationship between facial expression and speech modalities, and produce better discriminative emotional fusion features. Most classifiers produce decisions based on the confidence values that the sample belongs to, that is to say, the sample will be classified to belong to the class with the largest confidence value. The Softmax regression used in the proposal also complies with this rule. However, due to the fact that the emotional state information is ambiguous [19], this rule will lead to misclassification. Moreover, the feature level fusion used alone can not handle the situations well where the contributions of each modality data to emotion recognition are very imbalanced. As a result, the Fuzzy Broad Learning System (FBLS) [20] is introduced in the proposal which integrates the Takagi–Sugeno fuzzy model and broad learning system. Consequently, at the second stage, the FBLS is used to model the output confidence values of three Softmax regressions (Softmax regressions of facial expression, speech and fused feature) for final decision. To the best of our knowledge, in the previous studies, there are a few algorithms which not only consider internal relationship between facial expression and speech data, but also consider the ambiguity of emotional state information [21].

In this chapter, to extract representative facial expression and speech features and achieve effective modality fusion, Two-Stage Fuzzy Fusion based-Convolution Neural Network (TSFFCNN) is proposed for dynamic emotion recognition. The LBP-TOP and spectrogram are introduced first to extract low-level dynamic emotion features, so that the spatio-temporal information can be obtained from facial expression and speech data. To determine highly discriminative features for emotion recognition, two DCNNs are constructed to extract high-level emotion semantic features of facial expression and speech. Moreover, the TSFFS is developed by integrating CCA and FBLS, so as to consider the correlation and difference between different modal features. As a result, the TSFFCNN is formed. Not only it can extract discriminative emotion features which contain spatio-temporal information, but also effectively fuse facial expression and speech modalities at feature and decision level. Furthermore, the TSFFCNN is able to handle well the situations where the contributions of each modality data to emotion recognition are very imbalanced.

The chapter is organized as follows. Section 7.2 reviews the related works of the proposal. Section 7.3 introduces TSFFCNN for dynamic emotion recognition. Experimental results and analysis are presented in Sect. 7.4. Finally, Sect. 7.5 concludes the study.

7.2 Dynamic Emotion Feature Extraction

LBP-TOP [8] is employed to extract the low-level dynamic emotion features of facial expression. As shown in Fig. 7.2, the LBP-TOP regards the facial expression frames as the stack of XY planes in T-axis, namely, The XY plane contains the texture information of each frame, and the XT and YT planes contain the change of frames with time and space position. Figure 7.3 illustrates the process of low-level dynamic emotion feature completed by using LBP-TOP.

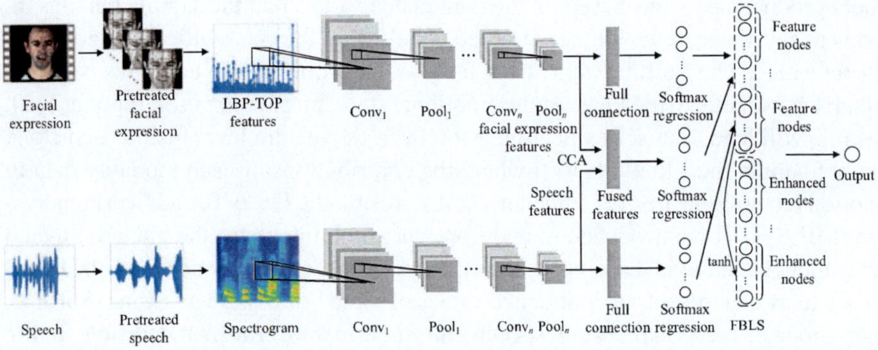

Fig. 7.1 The structure of two-stage fuzzy fusion based-deep convolution neural network

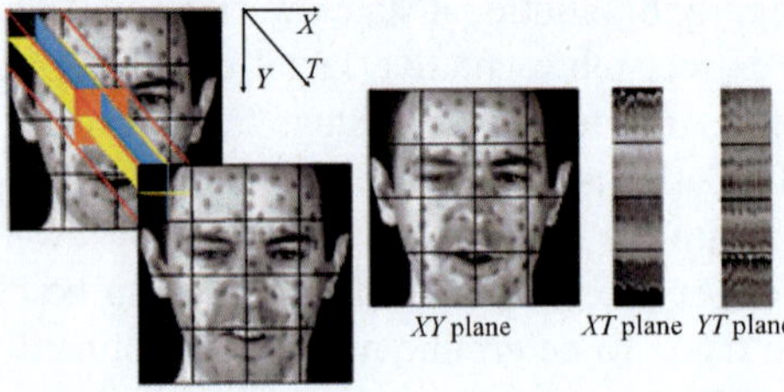

Fig. 7.2 Three planes in LBP-TOP

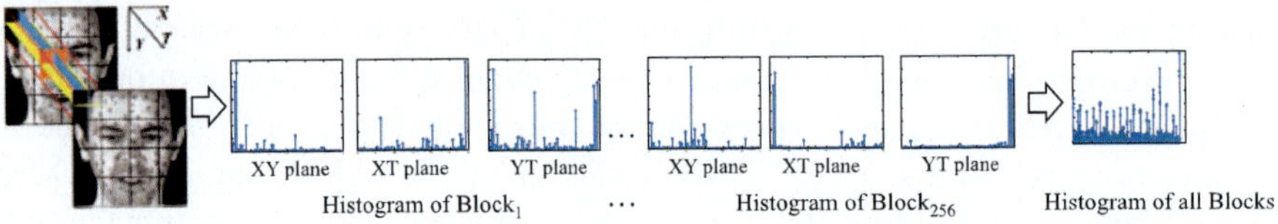

Fig. 7.3 The low-level dynamic emotion feature of facial expression by using LBP-TOP

Spectrogram represents the change of speech spectrum with time; its vertical axis denotes frequency, and its horizontal axis denotes time. The process for extracting spectrogram features is described as follows:

Firstly, the speech data y is processed by sub-frames to get framed speech sequence $y_d(n)$, $(d = 1, \ldots, D)$, which D denotes the number of frames.

Then, the framed speech sequence $y_d(n)$ is windowed to obtain the windowed speech data $y_{d,w}(n)$, namely

$$y_{d,w}(n) = y_d(n) * w(n) \tag{7.1}$$

where $w(n)$ denotes the Hanning window function expressed as

$$w(n) = 0.5(1 - \cos(2\pi \frac{n}{N})), 0 \le n \le N - 1 \tag{7.2}$$

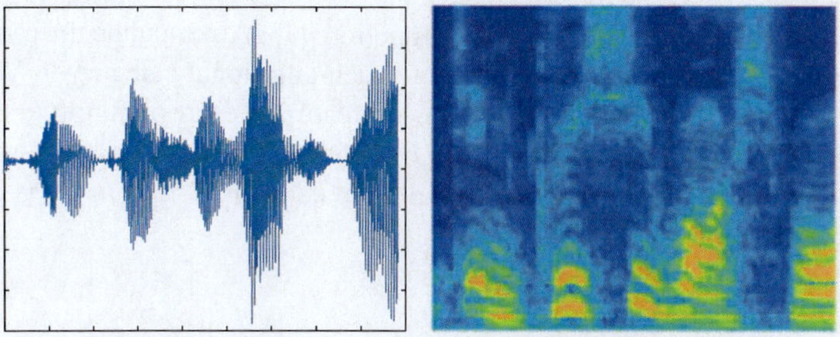

Fig. 7.4 Example spectrogram

Next, the Fast Fourier Transform (FFT) of $y_{d,w}(n)$ is calculated by (7.3), and the FFT coefficients $Y_d(k)$ are obtained.

$$Y_d(k) = \sum_{n=0}^{N-1} y_{d,w}(n) e^{-\frac{2\pi j}{N} kn}, 0 \le k \le N-1 \tag{7.3}$$

where the FFT coefficient matrix of $Y_d(k)$ is $Y = [Y_1, Y_2, \ldots, Y_D] \in \Re^{N \times D}$, and N represents the interval length of the FFT, D stands for the number of frames.

Finally, to smoothen out distributed data, a spectrogram B is produced

$$B = \log_{10}(Y + \varepsilon) \tag{7.4}$$

where ε is the regularization coefficient; see Fig. 7.4.

7.3 Deep Convolution Neural Network for Extracting High-Level Emotion Semantic Features

The DCNNs consist of a series of convolution layer, max-pooling layer, full connection layer, and Softmax regression in stack. The convolution layer moves the convolution filter along the vertical and horizontal directions of the input feature map to mine the local correlation information in the input feature map.

The activation function of RELU is used in the convolution layer. The max-pooling layer divides the input feature map into the rectangular pool area by max-pooling filters with the size of $m \times n$ and calculates the maximum value of each area to perform down-sampling through a pooling filter with the size of $m \times n$. The full connection layer in DCNN combines all of the features learned by the previous layers to identify the larger patterns. The input of Softmax regression are the features combined by full connection layer.

Softmax regression uses the hypothesis function $h_\theta(o)$ to calculate the probability that the input emotional feature o belongs to each emotional category j. When the probability value of the input emotional fusion feature o corresponding to the emotional category j is the largest among the J confidence values, the input emotional feature o is determined to belong to the emotional category j. The $h_\theta(o)$ is given by

$$h_\theta(o_i) = \begin{bmatrix} p(z_i = 1|o_i;\theta) \\ p(z_i = 2|o_i;\theta) \\ \vdots \\ p(z_i = J|o_i;\theta) \end{bmatrix} = \frac{1}{\sum_{j=1}^{J} e^{\theta_j^T o_i}} \begin{bmatrix} e^{\theta_1^T o_i} \\ e^{\theta_2^T o_i} \\ \vdots \\ e^{\theta_J^T o_i} \end{bmatrix} \tag{7.5}$$

where $z_i \in \{1, 2, \ldots, J\}$ is the labels of ith input data, J denotes that there are J class emotions.

The constructed DCNNs for facial expression and speech modalities are optimized separately by using the adaptive moment estimation (Adam) algorithm, and the high-level emotion semantic features F_h and B_h of facial expression and speech modalities are obtained from the full connection layer of the optimized DCNNs. Moreover, the confidence data C_F and C_B of facial expression and speech modalities can be obtained from the Softmax regressions.

7.4 Feature Fusion Based on Canonical Correlation Analysis

The CCA-based feature fusion is applied for fusing high-level emotion semantic features of facial expression and speech. Principal component analysis is used first to reduce the dimensionality of the high-level emotion semantic features F_h and B_h, and the dimension reduced features F_p and B_p of facial expression and speech are obtained. Then, the CCA is employed to fuse F_p and B_p, and the process is shown as follows:

The dimensionality of facial expression modal features $F_p \in \Re^{N\times U}$ and speech modal features $B_p \in \Re^{N\times V}$ are U and V, respectively, and the two sets of features have the same number of features N. The CCA maximizes the correlation between $\alpha^T F_p$ and $\beta^T B_p$ by looking for the projection vector α and β of the these two sets of feature data. To obtain α and β, the following criterion is defined

$$W(\alpha, \beta) = \frac{\alpha^T S_{FB}\beta}{\sqrt{\alpha^T S_{FF}\alpha \bullet \beta S_{BB}\beta^T}} \tag{7.6}$$

where S_{FF} and S_{BB} represent the covariance matrix of F_p and B_p, and S_{FB} represents the mutual covariance matrix of B_p and F_p.

In order to guarantee the uniqueness of the solution to (7.6), let $\alpha^T S_{FF}\alpha = 1$, $\beta S_{BB}\beta^T = 1$, and use the Lagrange multiplier method to transform the problem into the following two generalized expressions:

$$\begin{cases} S_{FB}S_{\mathrm{BB}}^{-1}S_{BF}\alpha = \lambda^2 S_{FF}\alpha \\ S_{BF}S_{FF}^{-1}S_{FB}\beta = \lambda^2 S_{BB}\beta \end{cases} \tag{7.7}$$

Let $M_{FF}=S_{FF}^{-1}S_{FB}S_{BB}^{-1}S_{BF}$, $M_{BB}=S_{BB}^{-1}S_{BF}S_{FF}^{-1}S_{FB}$, then (7.7) is converted into the following form

$$\begin{cases} M_{FF}\alpha = \lambda^2\alpha \\ M_{BB}\beta = \lambda^2\beta \end{cases} \tag{7.8}$$

We obtain the projection vectors α and β by solving (7.8), and obtain the discriminative features F_c and B_c.

$$\begin{cases} F_c = \alpha^T F_p \\ B_c = \beta^T B_p \end{cases}. \tag{7.9}$$

7.5 Decision Fusion Based on Fuzzy Broad Learning System

The FBLS is used to model the output confidence values of three Softmax regressions (Softmax regressions of facial expression, speech and fused feature) for

$$\begin{aligned} O_u^m &= \left(\sum_{i=1}^{I_m}\bar{\omega}_{ui}^m\left(\sum_{v=1}^{V}\rho_{i1}^m\mu_{iv}^m c_{uv}\right), \sum_{i=1}^{I_m}\bar{\omega}_{ui}^m\left(\sum_{v=1}^{V}\rho_{i2}^m\mu_{iv}^m c_{uv}\right), \ldots, \sum_{i=1}^{I_m}\bar{\omega}_{ui}^m\left(\sum_{v=1}^{V}\rho_{iJ}^m\mu_{iv}^m c_{uv}\right)\right) \\ &= \sum_{i=1}^{I_m}\bar{\omega}_{ui}^m\sum_{v=1}^{V}\mu_{iv}^m c_{uv}\left(\rho_{i1}^m, \rho_{i2}^m, \ldots, \rho_{iJ}^m\right) = \sum_{i=1}^{I_m}\bar{\omega}_{ui}^m g_{ui}^m\left(\rho_{i1}^m, \rho_{i2}^m, \ldots, \rho_{iJ}^m\right) \\ &= \left(\bar{\omega}_{u1}^m g_{u1}^m, \bar{\omega}_{u2}^m g_{u2}^m, \ldots, \bar{\omega}_{uI_m}^m g_{uI_m}^m\right)\begin{pmatrix} \rho_{11}^m & \cdots & \rho_{1J}^m \\ \vdots & \ddots & \vdots \\ \rho_{I_m1}^m & \cdots & \rho_{I_mJ}^m \end{pmatrix} \triangleq G_u^m\rho^m \end{aligned} \tag{7.10}$$

final decision, and the concrete structure of FBLS is illustrated in Fig. 7.5. The FBLS combines the Takagi–Sugeno fuzzy model and broad learning system, which not only exhibits the advantage of fast computation from broad learning system, but also has the mapping ability with continuous functions from Takagi–Sugeno fuzzy model. Broad learning system extends the neuron consisting of feature nodes and enhanced nodes extensively without deep superposition and uses pseudoinverse methods to calculate weights. Moreover, owing to the FBLS consisting of a series of Takagi–Sugeno fuzzy system, it is also able to benefit from ensemble learning.

Define M Takagi–Sugeno fuzzy sub-models and N groups enhanced nodes in the FBLS, and there are I_m fuzzy rules in the mth Takagi–Sugeno fuzzy sub-model. In

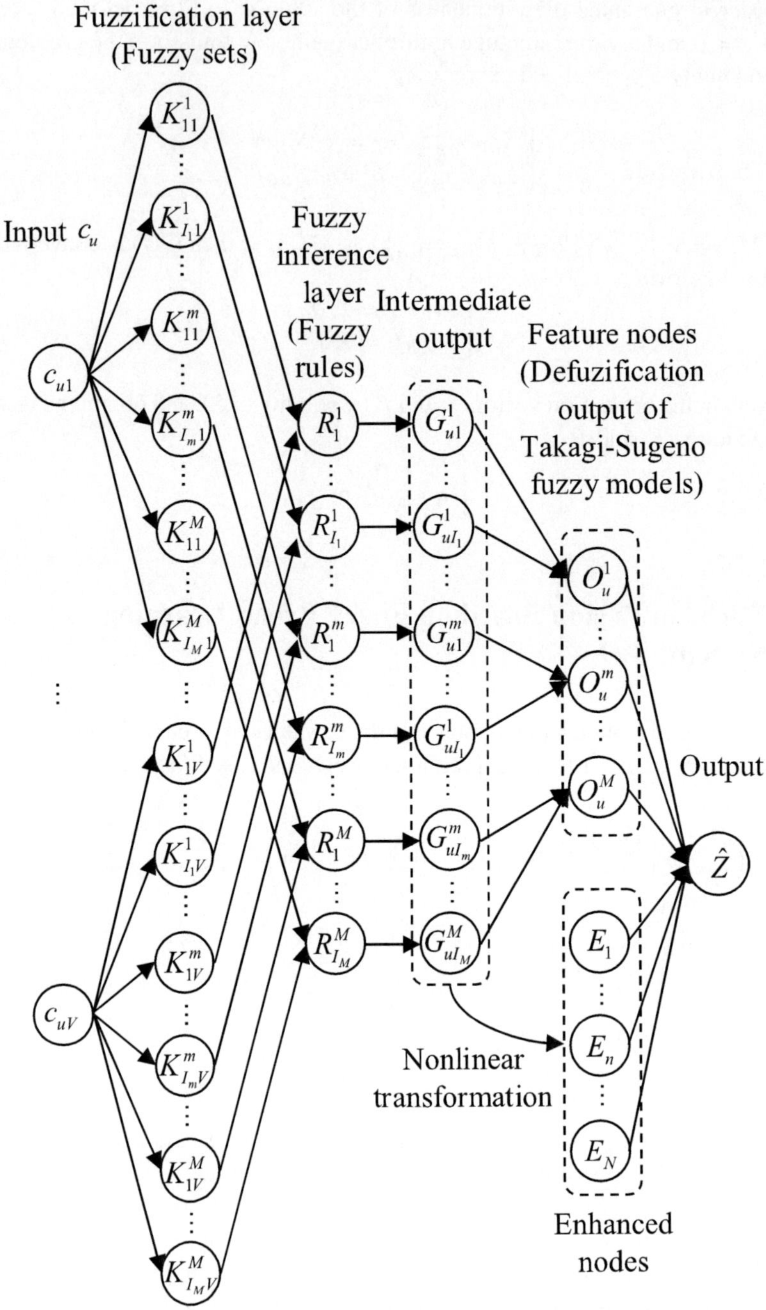

Fig. 7.5 Structure of FBLS

the mth Takagi–Sugeno fuzzy sub-models, the ith fuzzy rule R_i^m for the input data C is given as follows, where $C \in \Re^{U \times V}$ denotes U V-dimensional confidence data C obtained by Softmax regressions.

R_i^m: If c_{u1} is K_{i1}^m, and if c_{u2} is K_{i2}^m, ..., and if c_{uV} is K_{iV}^m, then $g_{ui}^m = \ell_{ui}^m(c_{u1}, c_{u2}, \ldots, c_{uV})$. $(i = 1, 2, \ldots, I_m)$

$\ell_{ui}^m(\cdot)$ is given by

$$g_{ui}^m = \ell_{ut}^m(c_{u1}, c_{u2}, \ldots, c_{uV}) = \sum_{v=1}^{V} \mu_{iv}^m c_{uv} \tag{7.11}$$

where g_{ui}^m denotes the output of fuzzy rule R_i^m, μ_{iv}^m denotes the consequential parameters of fuzzy rule R_i^m, and μ_{iv}^m is initialized randomly from [0, 1].

As shown in Fig. 7.5, in FBLS, the first layer is the fuzzification layer, where the input data C are transformed by Gaussian membership functions, which are given by

$$\xi_{iv}^m(c) = e^{-\left(\frac{c-\kappa_{iv}^m}{\sigma_{iv}^m}\right)^2} \tag{7.12}$$

where κ_{iv}^m denotes the position parameter, σ_{iv}^m denotes the scale parameter. And κ_{iv}^m is initialized via clustering centers I_m which is produced by using k-means clustering algorithm to the training set, and all σ_{iv}^m are set to 1.

The second layer is the fuzzy inference layer. Each node in this layer represents a fuzzy rule, which is obtained by multiplying the membership degree of the upper layer as the firing strength of the rule. The firing strength is given by

$$\omega_{ui}^m = \prod_{v=1}^{V} \xi_{iv}^m(c_{uv}) \tag{7.13}$$

The number of nodes in the third layer is the same as the one in the second layer, which represents the normalized value of the firing intensity of the fuzzy rules, and is given as

$$\bar{\omega}_{ui}^m = \frac{\omega_{ui}^m}{\sum_{i=1}^{I_m} \omega_{ui}^m} \tag{7.14}$$

In the fourth layer, the consequent part of the fuzzy rules is applied, the firing strength of the third layer is weighted as

$$G_{ui}^m = \bar{\omega}_{ui}^m g_{ui}^m \tag{7.15}$$

For the labels of one-hot encoding which corresponds to confidence data $C \in \Re^{U \times V}$ are given as $Z \in \Re^{U \times J}$, where J denotes J emotions. The defuzzification output $O_u^m \in \Re^J$ of the mth Takagi–Sugeno fuzzy sub-model for uth confidence data

c_u is given as (7.10), where $\rho^m \in \Re^{I_m \times J}$ demotes the coefficient of the consequence part for the mth Takagi–Sugeno fuzzy sub-model.

The output of mth Takagi–Sugeno fuzzy sub-model for confidence data C consists of all $O_u^m \in \Re^{U \times J}$ and is given as

$$\begin{aligned} O^m &= (O_1^m, O_2^m, \ldots, O_U^m)^T \\ &= (G_1^m \rho^m, G_2^m \rho^m, \ldots, G_U^m \rho^m)^T \\ &\triangleq G^m \rho^m \end{aligned} \tag{7.16}$$

The output $O \in \Re^{U \times J}$ of all Takagi–Sugeno fuzzy sub-models is given as

$$\begin{aligned} O &= \sum_{m=1}^{M} O^m = \sum_{m=1}^{M} G^m \rho^m \\ &= \left(G^1, G^2, \ldots, G^M\right)\left(\rho^1, \rho^2, \ldots, \rho^M\right) \triangleq G\rho \end{aligned} \tag{7.17}$$

To retain the information of the confidence data C as much as possible, the input of enhanced nodes don't consist of the defuzzification output of Takagi–Sugeno fuzzy sub-models, instead, the FBLS just concatenate all the G_{ui}^m to form a vector $G_u^m = (G_{u1}^m, G_{u2}^m, \ldots, G_{uI_m}^m)$, where G_u^m denotes the input data of enhanced nodes produced by the mth Takagi–Sugeno fuzzy sub-model for uth confidence data. And the input data of enhanced nodes produced by mth Takagi–Sugeno fuzzy sub-model for confidence data C consists of all G_u^m as $G^m = (G_1^m, G_2^m, \ldots, G_U^m)^T \in \Re^{U \times I_m}$, so that the input data of enhanced nodes produced by all Takagi–Sugeno fuzzy sub-models are given as $G = (G^1, G^2, \ldots, G^M) \in \Re^{U \times (I_1 + I_2 + \cdots + I_M)}$. Define P_n neurons in the nth group of enhanced node, and the output $E_n \in \Re^{U \times P_n}$ of nth group of enhanced node is given by

$$E_n = \hbar(G W_{e_n} + b_{e_n}) \tag{7.18}$$

where $n = 1, 2, \ldots, N$, W_{e_n} denotes the weights of nth group of enhanced nodes, b_{e_n} denotes the corresponding bias of W_{e_n}, and $\hbar(\cdot)$ denotes the tanh function. In this way, the output of all enhanced nodes are given as $E = (E_1, E_2, \ldots, E_N) \in \Re^{U \times (P_1 + P_2 + \ldots + P_N)}$.

At the top layer of FBLS, there are output $O \in \Re^{U \times J}$ of all Takagi–Sugeno fuzzy sub-models and the output of all enhanced nodes E. Define $W_c \in \Re^{(P_1 + P_2 + \ldots + P_N) \times J}$ as the connecting weights between enhanced layer and output layer, and weights between Takagi–Sugeno fuzzy models and output layer are 1. The output of FBLS is given by

$$\hat{Z} = O + E W_c = G\rho + E W_c = (G, E)(\rho, W_c)^T \triangleq (G, E) W_z \tag{7.19}$$

By using pseudoinverse, the W_z is obtained as

$$W_z = ((G, E)^T (G, E))^{-1} (G, E)^T Z. \tag{7.20}$$

7.6 Two-Stage Fuzzy Fusion Strategy

In the meanwhile, to consider the correlation and difference between different modal features, and deal with the situation where the contributions of each modality data to emotion recognition are very imbalanced, the TSFFS is developed by combining the CCA and FBLS. The steps for implementation of TSFFS are given as follows:

Step 1: Use principal component analysis to reduce the dimensionality of the high-level emotion semantic features F_h and B_h, and obtain the dimension reduced features F_p and B_p of facial expression and speech.

Step 2: Obtain the discriminative features F_c and B_c from F_p and B_p by using CCA, and connect F_c and B_c in series to get the fused feature $FB = [F_c, B_c]$.

Step 3: Use Softmax regression to recognize the fused feature FB, and the confidence data C_{FB} are obtained. Connect the confidence data C_F, C_B, and C_{FB} in series, namely, $C = [C_F, C_B, C_{FB}]$.

Step 4: Produce recognition results by using FBLS for final decision from confidence data C.

7.7 Experiments

The experiments on Surrey Audio-Visual Expressed Emotion (SAVEE) [28], eNTERFACE'05 [29], and Acted Facial Expressions In The Wild (AFEW) [30] databases are completed to verify the effectiveness of the proposal. The experiments are conducted on a computer with two dual core processers of both 2.4 GHz, 64 GB memory, a GPU of GeForce GTX 1080 Ti, and Windows 10 system. The software of the experiments is MATLAB R2018b. Apart from the training of DCNN which is run on the GPU, other parts are run on the CPU.

7.7.1 Data Setting

The SAVEE database consists of recordings coming from 4 male actors with an average age of 30 years in 7 basic emotions (angry, disgust, fear, neutral, happiness, sadness, and surprise).

It contains a total of 480 sets of video clips, the video sampling frequency is 44.1 kHZ, the frame rate is 60 fps.

The eNTERFACE'05 database contains 43 subjects from 14 different countries. Each subject listens to six stories in order to induce corresponding six basic emotions (anger, disgust, fear, happiness, sadness, and surprise). It contains a total of 1287 sets of video clips, the video sampling frequency is 48 kHZ, the frame rate is 25 fps. The AFEW database contains video clips collected from different movies which are believed to be close to the real-world environment. The data are divided into

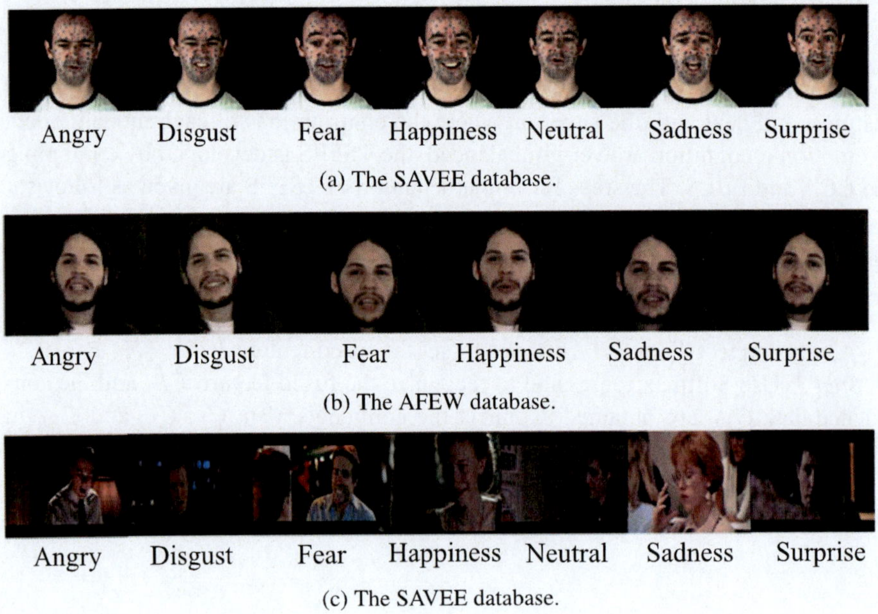

Fig. 7.6 Some samples in the SAVEE, eNTERFACE'05, and AFEW database

three parts, test set, training set, and validation set. Due to that, the labels of the test set cannot access to the public, the validation set and training set are used in the experiments. In total, 1107 sets of video clips are used in the experiments which cover the seven basic expressions, the video sampling frequency is 48 kHZ, the frame rate is 25 fps. Figure 7.6 includes some the samples in the SAVEE, eNTERFACE'05, and AFEW database. The ten-fold cross validation method is used in these experiments.

7.7.2 *Experiments for Hyperparameters*

There are several hyperparameters which have significate influence on the performance of the proposal, i.e., the number of key frames and the number of blocks in facial expression modality; the structure of DCNNs in both facial expression and speech modalities; the number of Takagi–Sugeno fuzzy sub-model, rules in the Takagi–Sugeno fuzzy model, and enhanced nodes in FBLS. Therefore, a series of experiments are conducted for these hyperparameters by using ten-fold cross validation on eNTERFACE'05 database.

Figure 7.7 shows the experimental results with the change of the number of key frames. In the experiments, the DCNN is used for facial expression recognition whose structure is shown in Table 7.1, and the block segmentation is not conducted, namely, when extracting the low-level dynamic emotion feature by using LBP-TOP,

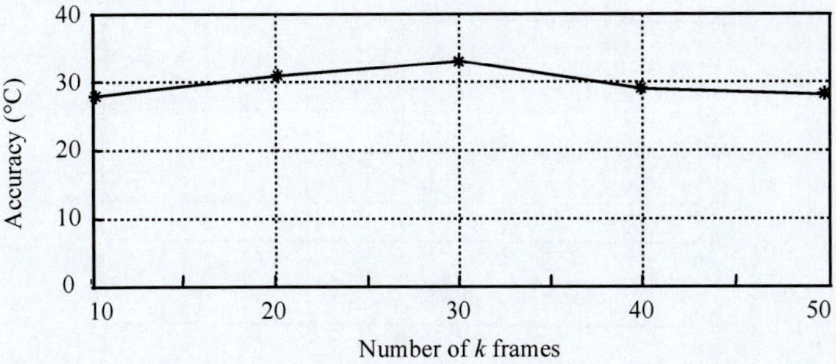

Fig. 7.7 Experimental results with the change of the number of key frame for eNTERFACE'05 database

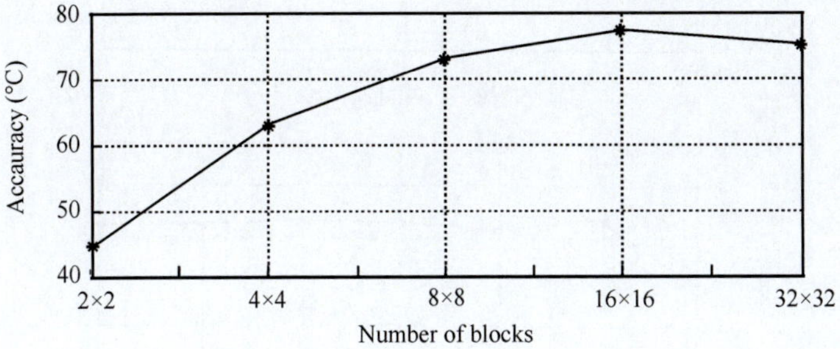

Fig. 7.8 Experimental results with the change of the number of block for eNTERFACE'05 database

the whole image of each frame are used. As illustrated in Fig. 7.7, the experimental accuracy is the highest when the number of key frames is 30. When the number of key frames is less than 30, one is not able to extract enough information for dynamic emotion recognition. When the number of key frames is more than 30, it will cause information redundancy, which is not conducive to dynamic emotion recognition. In this case, the number of key frames is set to 30 as a sound choice.

The experimental results concerning the change of the number of blocks are shown in Fig. 7.8. In these experiments, the number of key frames is 30, and the DCNN is also used for facial expression recognition whose structure in shown is Table 7.1. It can be seen from Fig. 7.8, when the number of blocks is 16×16, the accuracy is the highest. When the number of blocks is less than 16×16, it can not grasp the details of facial expression changes for dynamic emotion recognition. When the number of blocks is more than 16×16, it will also cause information redundancy. Hence, the number of blocks is set to 16×16 in the proposal.

In the popular architecture of DCNNs (e.g., VGG16, VGG19, GoogLeNet, DenseNet, etc.), the size of filter in the convolution layer is usually set as 3×3, and the number of convolutional filter increases with the number of layers. Inspired by this, the height and width of convolution filter are designed on the basis of 3. Considering that the size of low-level dynamic emotion feature for facial expression and speech are 128×98 and 45312×1 where the height and width of features are

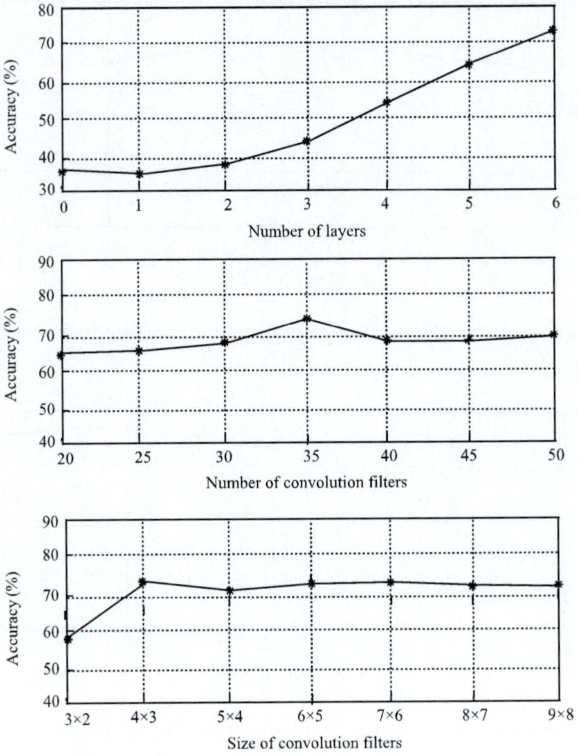

Fig. 7.9 Experimental results of the DCNN of facial expression modality

different, thus, the size of filter in the convolution layer for facial expression and speech are set to 4×3 and 3×1.

To establish appropriate architecture of DCNNs for facial expression and speech modalities, a series of experiments are carried out with the change of the number of layers and convolution filters, and the size of convolution filters. To illustrate the experimental results with the change of the number of layers more intuitively, we define a series numbers which are given in the abscissa for the "Number of layers" in Fig. 7.9 to denote the structure of DCNNs used in the experiments. The number 0 denotes that low-level dynamic emotion features of facial expression modality or speech modality are input to the Softmax regression for recognition directly. The number 1–5 denotes that how many convolution layers and max-pooling layers are stacked in the DCNN, namely, the 1 denotes a convolution layer and a max-pooling layer, the 2 denotes two convolution layer and two max-pooling layer, and so on; and followed with the convolution layers and max-pooling layers are full connection layer and Softmax regression. And the number of filters increases with the increase of convolution layers, i.e., for the DCNN of facial expression modality, the first convolution layer has 100 filters, the second convolution layer has 200 filters, the third convolution layer has 300 filters, and so on; for the DCNN of speech modality, the first convolution layer has 35 filters, the second convolution layer has 70 filters,

Table 7.1 Structural parameters of the convolutional neural network used for facial expression modality

Layer no.	Layer type	Filter size (Stride)	Filter no.
1	Convolution	3×1(1)	100
2	Max-pooling	3×1(2)	\

Table 7.2 Structural parameters of the convolutional neural network for speech modality

Layer no.	Layer type	Filter size (Stride)	Filter No.
1	Convolution	4×3 (1)	35
2	Max-pooling	4×3 (2)	\
3	Convolution	4×3 (1)	35×2
4	Max-pooling	4×3 (2)	\
5	Convolution	4×3 (1)	35×3
6	Max-pooling	4×3 (2)	\
7	Convolution	4×3 (1)	35×4
8	Convolution	4×3 (1)	35×5
9	Max-pooling	1×13 (1)	\

the third convolution layer has 105 filters, and so on. Moreover, for the DCNNs of facial expression modality, the size of convolution filters and max-pooling filters are the size of filters are set as 3×1 with stride 1; for the DCNNs of speech modality, the size of convolution filters and max-pooling filters are the size of filters are set as 3×4 with stride 1. For speech modality, the number 6 denotes the structure of DCNN presented in Table 7.2; for facial expression modality, the number 6 also denotes the structure of DCNN presented in Table 7.2, but the size of convolution filters and max-pooling filters are the size of filters are set as 3×1 with stride 1.

It can be seen from Fig. 7.8, with the increase of convolution layer and max-pooling layer, the performance of the DCNN for facial expression becomes worse, in contrast, the performance of the DCNN for speech becomes better. This may be due to that the speech emotion recognition is not only affected by the expression, sound quality, and accent of each person, but also affected by different speech content, which makes the speech emotion data more complicated and more difficult to recognize compared with facial expression data. Therefore, the structure of the DCNN for speech modality is more complicated than that of facial expression modality. According to Fig. 7.9, when the layer is 1 for the DCNN of facial expression modality and the layer is 6 for the DCNN of speech modality, the performance of the DCNNs is best, so that the number of layers for the DCNNs of facial expression and speech modalities are set as presented in Tables 7.1 and 7.2.

In the experiments with the change of the number of convolution filters, except for the number of convolution filters, the other parameters of the DCNNs are set as presented in Tables 7.1 and 7.2. In Fig. 7.10, the abscissa for the "Number of

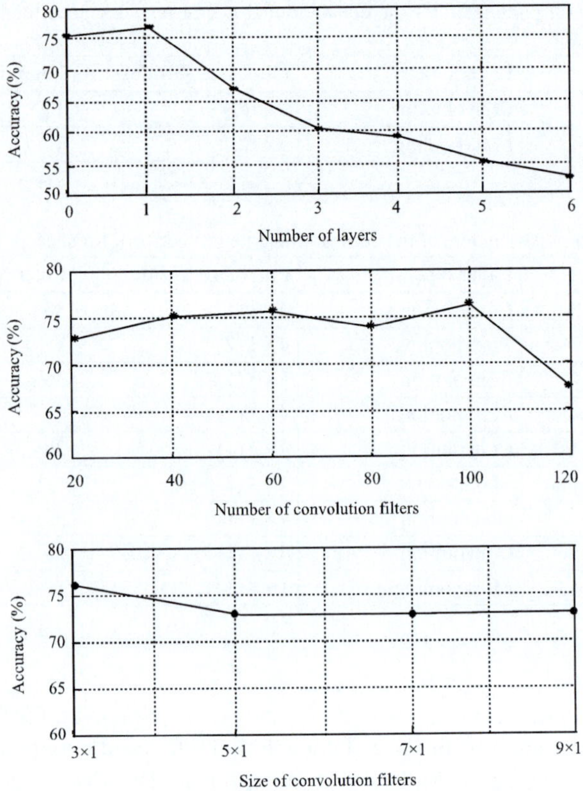

Fig. 7.10 Experimental results of the DCNN of speech modality

convolution filters" denotes the number of convolution filters in the first convolution layer. For the DCNN of speech modality, the number of convolution filters of the followed convolution layer increases with the number of layers by a corresponding multiple as presented in Table 7.2. That is to say, if the number of convolution filters in the first convolution layer is 35, the number of convolution filters in the second, third, fourth, and fifth convolution layer is 35×2, 35×3, 35×4, and 35×5, respectively. As shown in Fig. 7.10, when the number of filters is 100, the DCNN of facial expression modality has the best performance, and the number of filters in the first convolution layer is 35, the DCNN of speech modality has highest accuracy, so the number of filters in the DCNN of facial expression modality is 100, the number of filters in the first convolution layer for the DCNN of facial expression modality is 35.

In the experiments dealing with the change of the size of convolution filters, except for the size of convolution filters, the other parameters of the DCNNs are set as presented in Tables 7.1 and 7.2. As shown in Fig. 7.10, when the size of convolution filters is 4×3 and 3×1 in the DCNNs of facial expression and speech modality, the DCNNs have the best performance. Hence, the the size of convolution filters is set as 4×3 and 3×1 in the DCNNs of facial expression and speech modalities.

Moreover, to evaluate the impact of the number of rules, fuzzy systems, and enhanced nodes on the FBLS, the experiments with the change of these parameters are conducted. Figure 7.10 gives the experimental results with the change of the number of rules and fuzzy systems, where the number of enhanced nodes is fixed as 8. As shown in Fig. 7.10, too many rules and fuzzy systems may cause overfitting, and the accuracy will decrease, so that the number of rules and fuzzy systems are set to 4 and 3 according to the experimental results. Figure 7.12 illustrates the experimental results with the change of the number of enhanced nodes, where the number of rules and fuzzy systems are fixed as 4 and 3. According to Fig. 7.12, when the number of enhanced nodes is 8, the accuracy is the highest. Too many or too little enhanced nodes can not preserve the characteristic of inputs data, and it will deteriorate the performance of FBLS. As a result, the number of enhanced nodes is set to 8.

7.7.3 Experimental Results and Analysis

The experimental results on SAVEE, eNTERFACE'05, and AFEW databases by using the proposed TSFFCNN are shown in Table 7.3. The experiments were carried out by using ten-fold cross validation. As shown in Table 7.3, the average recognition accuracies on SAVEE database with the single modality of facial expression and speech, and both modalities are 97.71%, 62.54%, and 99.79%, respectively; the average recognition accuracies on eNTERFACE'05 database with the single modality of facial expression and speech, and both modalities are 76.38%, 73.27%, and 90.82%, respectively; the average recognition accuracies on AFEW database with the single modality of facial expression and speech, and both modalities are 35.70%, 23.92%, and 50.28%, respectively. It can be seen from Table 7.3 that the recognition accuracies of facial expression modality are higher than that of speech modality, especially the experimental results on SAVEE database. This may be due to that the speech data are more complicated than facial expression data, it is hard to extract discriminative features for dynamic emotion recognition, particularly when the number of samples is small. As for the experimental results on AFEW database, the facial expression data include unconstrained facial expressions, different head postures, a large age range, occlusion, different focal points, different facial partial resolutions, and near-real-world lighting; the speech data contain the background music and other noise in the real-world environment, some of the data even only contain the background music. In this case, this database is much more complex than SAVEE and eNTERFACE'05 databases, while the scale of AFEW database is similar as eNTERFACE'05 database, so the accuracies of AFEW database are relatively lower.

Figures 7.13, 7.14, and 7.15 show the confusion matrices obtained the SAVEE, eNTERFACE'05, and AFEW database. As it is shown in Fig. 7.13, other than the emotion of sadness whose recognition accuracy is 96.67%, the recognition accuracies are 100%, and the confusion appears between sadness and neutral. There are samples of sadness emotion which is misclassified as neutral, this is due to that the confidence

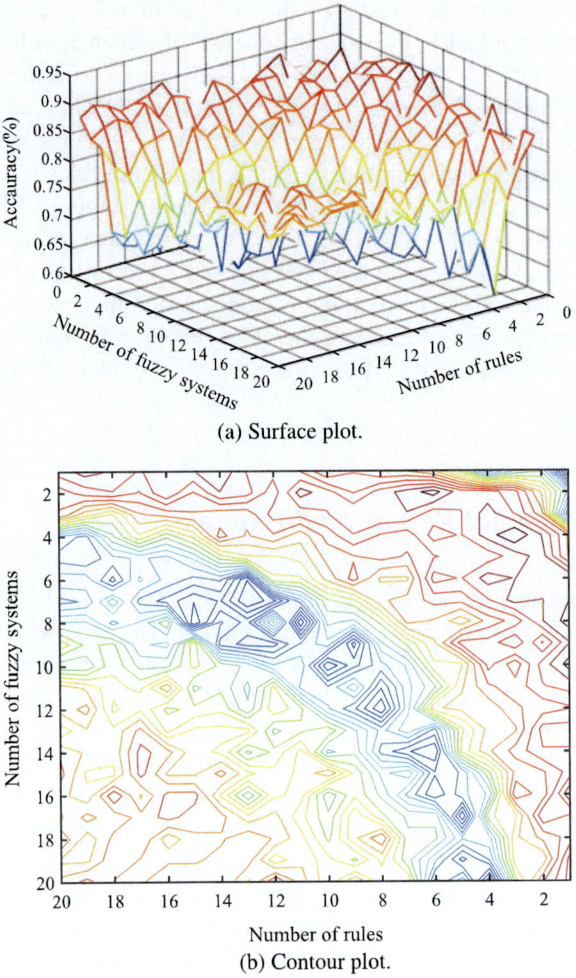

Fig. 7.11 The experimental results showing the change of the number of rules and fuzzy systems for eNTERFACE'05 database

Table 7.3 The experimental results on benchmark databases by using TSFFCNN (Mean±Std)

Database	Facial expression	Speech	Both
SAVEE	97.71%±2.07	62.54%± 15.18	99.79%±0.66
eNTERFACE'05	76.38%±5.35	73.27%±3.32	90.82%±2.35
AFEW	35.70%±2.55	23.92%±2.54	50.28%±1.73

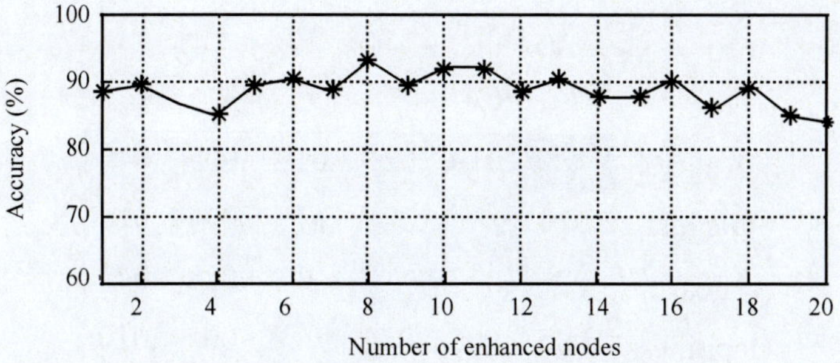

Fig. 7.12 The experimental results with the change of the number of enhanced nodes for eNTERFACE'05 database

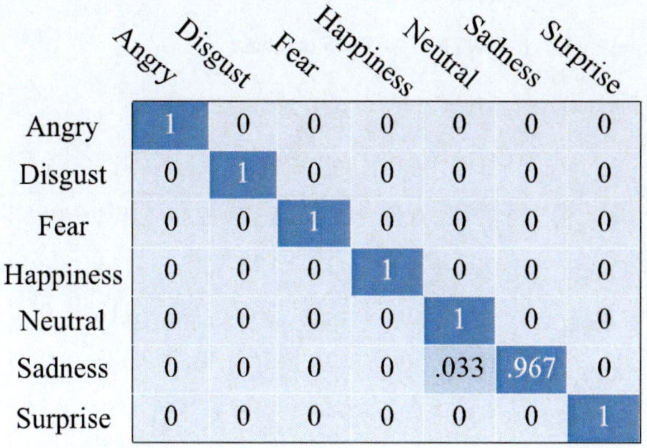

Fig. 7.13 Confusion matrix for SAVEE database

data of Softmax regressions are have high confidence in other categories, but low confidence in sadness, which makes it difficult to make correct decisions in the final decision-making stage. According to Fig. 7.14, the recognition accuracies of all emotions are higher than 86%, the emotion of happiness achieves the highest accuracy with 95.3%, the lowest accuracy is fear with 86.4%, and the relative higher confusion appears between angry and surprise, fear and surprise. It can be seen from Fig. 7.15, the recognition accuracies of happiness and neutral are 80.6% and 70.3%, respectively. There are relatively high confusion levels between fear and happiness, fear and neutral, sadness and neutral, disgust and neutral.

Moreover, the total training time and test time of the proposed TSFFCNN on the experiments of each fold on eNTERFACE'05 database are about 811.31s and 6.39s, respectively. Therefore, the training time and test time for each sample is about 0.70s and 0.050s, respectively. It can be seen that the developed approach can meet the real-time requirements.

To verify the validity of the proposed TSFFCNN, the TSFFCNN is compared with other approaches which are shown in Table 7.4, i.e., ISLA [26], DEW [23], HDM [4], RBML [3], SKRRR [24], DMCCA [25], MFF [22], FAMNN [27], and

	Angry	Disgust	Fear	Happiness	Sadness	Surprise
Angry	.879	.019	.027	.019	.009	.046
Disgust	.014	.907	.023	.019	.024	.014
Fear	.024	.037	.864	0	.028	.047
Happiness	.005	.005	.024	.953	.005	.010
Sadness	.010	.009	.033	0	.939	.009
Surprise	.028	.014	.037	.014	.037	.869

Fig. 7.14 Confusion matrix for eNTERFACE'05 database

	Angry	Disgust	Fear	Happiness	Neutral	Sadness	Surprise
Angry	.421	.037	.039	.132	.224	.091	.056
Disgust	.222	.139	.028	.194	.250	.111	.056
Fear	.205	0	.170	.267	.240	.118	0
Happiness	.042	0	.028	.806	.069	.055	0
Neutral	.030	.015	.044	.119	.703	.059	.030
Sadness	.119	.021	.041	.146	.266	.346	.061
Surprise	.111	0	.028	.222	.139	.056	.444

Fig. 7.15 Confusion matrix for AFEW database

CCA based CNN (CCACNN). As illustrated in Table 7.4, the performance of TSFFCNN is superior to CCACNN which just uses the CCA to fuse the high-level emotion semantic features of facial expression and speech modalities by feature-based fusion, the accuracies of TSFFCNN obtained on SAVEE, eNTERFACE'05, and AFEW databases are 99.79%, 90.82% and 50.28%, the accuracies of CCACNN obtained on SAVEE, eNTERFACE'05, and AFEW databases are 96.88%, 86.24%, 42.56%. Moreover, according to the experimental results on SAVEE database which is shown in Tables 7.3 and 7.4, when the contributions of each modality data to emotion recognition are very imbalanced, the recognition accuracy by using CCACNN to fuse features is lower than that of single modality. However, the proposed TSFFCNN can handle this situation well. And performance of the proposal is also better than that of the-state-of-art approaches as shown in Table 7.4, i.e., deep learning-based feature extraction and multimodal fusion methods DEW [23] and HDM [4], feature-based fusion methods SKRRR [24], DMCCA [25] and MFF [22], decision-based

Table 7.4 Comparative analysis the existing approaches obtained on benchmark databases

Database	Method	Accuracy
SAVEE	ISLA [26]	86.01%
	FAMNN [27]	98.25%
	CCACNN	96.88%±2.99
	TSFFCNN	99.79%±0.66
eNTERFACE'05	DEW [23]	85.69%
	HDM [4]	85.97%
	RBML [3]	86.67%
	DMCCA [25]	87.00%
	SKRRR [24]	87.40%
	CCACNN	86.24%±3.61
	TSFFCNN	90.82%±2.35
AFEW	MFF [22]	45.2%
	SKRRR [24]	47.00%
	CCACNN	42.56%±2.08
	TSFFCNN	50.28%±1.73

fusion methods ISLA [26], and hybrid fusion methods RBML [3] and FAMNN [27]. The accuracies of ISLA [26] and FAMNN [27] on SAVEE database are 86.01% and 98.25%, respectively. The accuracies of DEW [23], HDM [4], RMBL [3], DMCCA [25], and SKRRR [24] on eNTERFACE'05 database are 85.69%, 85.97%, 86.67%, 87.00%, and 87.40%, respectively. The accuracies of SKRRR [24] and MFF [22] on AFEW database are 45.2% and 47.00%, respectively. It follows that the proposed TSFFCNN not only can extract discriminative emotion features which contains spatio-temporal information, but also can effectively fuse facial expression and speech modalities by hybrid fusion.

Although the results presented in [2] are higher than that of the proposal, which achieves 100% on SAVEE database and 98.73% on eNTERFACE'05 database, the computational cost of the algorithm in [2] is much higher than that of TSFFCNN. This is because all facial expression frames are used in facial expression and recognized by GoogLeNet which is pre-trained on a filtered mix of the Imbd-Wiki and Audience databases, where the number of video frames in each video in SAVEE and eNTERFACE'05 databases can reach hundreds of frames. As a result, the training process of GoogLeNet can be very time-consuming. The experimental results of GoogLeNet for facial expression on SAVEE (97.50%) and eNTERFACE'05 (63.56%) databases presented in [2] are lower than that of TSFFCNN (97.71% and 76.38%), the statistical methods for the experimental results in [2] are different from ours. The expression of emotion usually includes the stage of neutral-apex-neutral, so if each frame is recognized separately, those frames with less salient emotion are easily misclassified, which leads to lower recognition accuracy. But when the confidence values of all these frames and other features are combined to make decisions, the confidence values of these video frames can be regarded as features with spatio-temporal information, and contribute to the final performance of emotion recognition. The overall performance on SAVEE and eNTERFACE'05 databases presented in [2] is better than that of ours.

7.8 Summary

In order to extract representative facial expression and speech features and achieve effective modality fusion, a TSFFCNN is proposed for dynamic emotion recognition which not only can extract discriminative emotion features which contains spatio-temporal information, but also effectively fuse facial expression and speech modalities from feature and decision fusion level.

In the proposal, the data preprocessing of facial expression and speech data is conducted first on the two modalities data. Then, the LBP-TOP and spectrogram are employed to low-level extract dynamic emotion features of facial expression and speech data, so that the spatio-temporal information can be obtained from facial expression and speech data. Next, two DCNNs are constructed for extracting high-level semantic emotion features from LBP-TOP and spectrogram features, in this way, the more discriminative features are gained. Finally, the TSFFS is developed to fuse the confidences of different modalities, where the CCA is applied to fuse the high-level emotion semantic features of facial expression and speech, and the FBLS is utilize to fuse the confidence values of facial expression, speech, and fused features of facial expression and speech for final decision. In this regard, the correlation and difference between different modal features are considered, as well as handle the ambiguity of emotional state information.

The experiments reported on benchmark databases are carried out to verify the validity of the proposed method, and the experimental results on SAVEE, eNTERFACE'05 and AFEW databases by using the proposed TSFFCNN are 99.79%, 90.82%, and 50.28%, which are higher than those of the existing methods on SAVEE, eNTERFACE'05 and AFEW databases, e.g., deep learning-based feature extraction and modality fusion method HDM, and hybrid fusion based method RBML.

In the future, the emotion recognition from more modalities will be considered, i.e., facial expression, speech, body gesture, and electroencephalogram, which is close to real world scenarios. The end-to-end approach which can learn discriminative features from different modalities directly without extracting low-level features will be further explored. Moreover, the emotional intention understanding becomes more important for robots when interacting with humans, so that the emotional intention understanding will be explored as well.

References

1. L. Pessoa, How do emotion and motivation direct executive control. Trends Cognit. Sci. **13**(1), 160–166 (2009)
2. F. Noroozi, M. Marjanovic, A. Njegus, S. Escalera, G. Anbarjafari, Audio-visual emotion recognition in video clips. IEEE Trans. Affect. Comput. **10**(1), 60–75 (2019)
3. K.P. Seng, L. Ang, C.S. Ooi, A combined rule-based & machine learning audio-visual emotion recognition approach. IEEE Trans. Affect. Comput. **9**(1), 3–13 (2018)
4. S. Zhang, S. Zhang, T. Huang, W. Gao, Q. Tian, Learning affective features with a hybrid deep model for audio-visual emotion recognition. IEEE Trans. Circuits Syst. Video Technol. **28**(10), 3030–3043 (2018)
5. A. Majumder, L. Behera, V.K. Subramanian, Automatic facial expression recognition system using deep network-based data fusion. IEEE Trans. Cybern. **48**(1), 103–114 (2018)
6. M. Emambakhsh, A. Evans, Nasal patches and curves for expression-robust 3d face recognition. IEEE Trans. Pattern Anal. Mach. Intell. **39**(5), 995–1007 (2017)
7. L. Chen, M. Zhou, M. Wu, J. She, Z. Liu, F. Dong, K. Hirota, Three-layer weighted fuzzy support vector regression for emotional intention understanding in human-robot interaction. IEEE Trans. Fuzzy Syst. **26**(5), 2524–2538 (2018)
8. G. Zhao, M. Pietikainen, Dynamic texture recognition using local binary patterns with an application to facial expressions. IEEE Trans. Pattern Anal. Mach. Intell. **29**(6), 915–928 (2007)
9. H. Hermansky, Coding and decoding of messages in human speech communication: implications for machine recognition of speech. Speech Commun. **106**, 112–117 (2019)
10. H.M. Fayek, M. Lech, L. Cavedon, Evaluating deep learning architectures for speech emotion recognition. Neural Netw. **92**, 60–68 (2017)
11. P. Jiang, H. Fu, H. Tao, P. Lei, L. Zhao, Parallelized convolutional recurrent neural network with spectral features for speech emotion recognition. IEEE Access **7**(90), 368–377 (2019)
12. L. Chen, W. Su, Y. Feng, M. Wu, J. She, K. Hirota, Two-layer fuzzy multiple random forest for speech emotion recognition in human-robot interaction. Inf. Sci. **509**, 150–163 (2020)
13. S. Zhang, S. Zhang, T. Huang, W. Gao, Speech emotion recognition using deep convolutional neural network and discriminant temporal pyramid Matching. IEEE Trans. Multimedia **20**(6), 1576–1590 (2018)
14. Y. LeCun, Y. Bengio, G. Hinton, Deep learning. Nature **521**, 436–444 (2015)
15. M. Wu, W. Su, L. Chen, Z. Liu, W. Cao, K. Hirota, Weight-adapted convolution neural network for facial expression recognition in human-robot Interaction. IEEE Trans. Syst. Man Cybern. Syst. (2019) https://doi.org/10.1109/TSMC.2019.2897330
16. S.K. D'mello, J. Kory, A review and meta-analysis of multimodal affect detection systems. ACM Comput. Surv. **47**(3), 1–36 (2015)
17. S. Poria, E. Cambria, R. Bajpai, A. Hussain, A review of affective computing: from unimodal analysis to multimodal fusion. Inf. Fus. **37**, 98–125 (2017)
18. T. Baltrušaitis, C. Ahuja, L. Morency, Multimodal machine learning: a survey and taxonomy. IEEE Trans. Pattern Anal. Mach. Intell. **41**(2), 423–443 (2019)
19. K.Y. Chan, U. Engelke, Varying spread fuzzy regression for affective quality estimation. IEEE Trans. Fuzzy Syst. **25**(3), 594–613 (2017)
20. S. Feng, C.L.P. Chen, Fuzzy broad learning system: a novel neuro-fuzzy model for regression and classification. IEEE Trans. Cybern. **50**(2), 414–424 (2020)
21. T.L. Nguyen, S. Kavuri, M. Lee, A multimodal convolutional neuro-fuzzy network for emotion understanding of movie clips. Neural Netw. **118**, 208–219 (2019)
22. J. Chen, Z. Chen, Z. Chi, H. Fu, Facial expression recognition in video with multiple feature fusion. IEEE Trans. Affect. Comput. **9**(1), 38–50 (2018)
23. Y. Ma, Y. Hao, M. Chen, J. Chen, P. Lu, A. Košir, Audio-visual emotion fusion (avef): a deep efficient weighted approach. Inf. Fus. **46**, 184–192 (2019)
24. J. Yan, W. Zheng, Q. Xu, G. Lu, H. Li, B. Wang, Sparse kernel reduced-rank regression for bimodal emotion recognition from facial expression and speech. IEEE Trans. Multimedia **18**(7), 1319–1329 (2016)

25. L. Gao, L. Qi, E. Chen, L. Guan, Discriminative multiple canonical correlation analysis for information fusion. IEEE Trans. Image Process. **27**(4), 1951–1965 (2018)
26. Y. Kim, E.M. Provost, Isla: temporal segmentation and labeling for audio-visual emotion recognition. IEEE Trans. Affect. Comput. **10**(2), 196–208 (2019)
27. D. Gharavian, M. Bejani, M. Sheikhan, Audio-visual emotion recognition using fcbf feature selection method and particle swarm optimization for fuzzy artmap neural networks. Multimedia Tools Appl. **76**(2), 2331–2352 (2017)
28. S. Haq, P. Jackson, J. Edge, Audio-visual feature selection and reduction for emotion classification, in *Proceedings of international conference on auditory-visual speech processing*, pp. 185–190 (2008)
29. O. Martin, I. Kotsia, B. Macq, I. Pitas, The enterface' 05 audio-visual emotion database, in *Proceedings of 22nd International Conference on Data Engineering Workshops (ICDEW'06)*, pp. 1–8 (2006)
30. A. Dhall, R. Goecke, S. Lucey, T. Gedeon, Collecting large, richly annotated facial-expression databases from movies. IEEE MultiMedia **19**(3), 34–41 (2012)

Chapter 8
Multi-support Vector Machine Based Dempster-Shafer Theory for Gesture Intention Understanding

The Dempster-Shafer (D-S) theory based on multi-SVM to deal with multimodal gesture images for intention understanding is proposed, in which the Sparse Coding (SC) based Speeded-Up Robust Features (SURF) are used for feature extraction of depth and RGB image. Aiming at the problems of the small sample, high dimensionality and feature redundancy for image data, we use the SURF algorithm to extract the features of the original image, and then perform their Sparse Coding, which means that the image is subjected to two-dimensional feature reduction. The dimensionally reduced gesture features are used by the multi-SVM for classification. A fusion framework based on D-S evidence theory is constructed to deal with the recognition of depth and RGB image to realise the gesture intention understanding.

8.1 Introduction

In recent years, research on human gesture intention understanding has drawn considerable attention [1] for its broad application prospects [2–4]. Researchers in intelligent human-robot interaction (HRI) have been committed to empower robots to perceive [5], recognize [6], and analyze human behaviors [7], so as to understand human intentions and emotions.

Some specific human gestures often have specific behavioral intentions. There has been related research on the gesture intention understanding. In this field, Ahmad et al. [8] judged human driving intentions by locating and tracking people's eyes and gestures, which is beneficial to accomplish Human-robot interaction tasks. Aparna et al. [9] identify Indian classical dance to understand human emotions and intentions by means of deep learning methods and some emotion model. Gesture intention understanding refers to the whole process of collecting human gestures, extracting

L. Chen et al., *Emotion Recognition and Understanding for Emotional Human-Robot Interaction Systems*, Studies in Computational Intelligence 926,
https://doi.org/10.1007/978-3-030-61577-2_8

gesture features to translating into semantic intents [10]. Many studies use RGB-D sensors to collect gestures, the study reported in [11] realized the problem of human behavior detection based on the RGB-D camera in the home environment. Su et al. [12] developed a system based on Kinect to help disabled patients in the family environment.

In terms of gesture feature extraction, common feature extraction algorithms such as Scale-Invariant Feature Transform (SIFT) [13], Speeded-up Robust Feature (SURF) algorithm [14], FAST [15] algorithm, etc. belong to the low-level image representation method. However, high-level gesture image representation is more significant for image understanding. Since the mapping of such low-level features as SURF and classification objects is a very complex nonlinear function, it is difficult to directly construct the model for training, and achieve high classification performance [16–18]. Therefore, it is necessary to build a salient high-level feature of the data. This arises as a first motivating factor behind the research reported in this study. Many machine learning methods are used to learn high-level representations of images, such as Principal Component Analysis, Sparse Coding, and low-rank representation [19–21], etc. Sparsity is a characteristic of an image itself [22]. As an up-and-coming signal processing technology, sparse coding is an unsupervised learning algorithm which is used to find a set of super-complete basis vectors to represent the input vector as a linear combination of these base vectors. Yang et al. [23] proposed a Sparse Coding Spatial Pyramid Matching (ScSPM) method which is a combination of sparse coding and linear kernel for image classification.

Research shows that the fusion of multiple methods is the promising development direction of future gesture recognition [24]. The underlying data or multimodal data are heterogeneous. Numerous data processing methods are integrated or adopted by multimodal information processing systems to complete the identification of the same target [25–27]. The multimodal data fusion method includes pixel level, feature level, and decision level fusion, as well as the score level and rank level fusion, which is used in multi-biometric systems when the individual matcher's output is a ranking of the "candidates" in the template database sorted in a decreasing order (or, an increasing order in appropriate cases) of match scores [28]. The D-S evidence theory proposed by Dempster and Shafer can effectively solve the problem of data inaccuracy and uncertainty [29] in the absence of prior probability, and is widely used in the field of information fusion [30–32].

In this chapter, a fusion framework is introduced that utilizes data coming from RGB and depth sensors to realize gesture intention understanding [33]. The Sparse coding algorithm based on SURF feature is used for gesture feature extraction, and multi-class linear SVM and the D-S evidence theory are used for feature classification as well as gesture intention understanding. To verify the effectiveness of the proposal, experiments on two RGB-D datasets (CGD2011 and CAD-60) are conducted. The originality of this chapter lies in the use of a fusion framework for behavior recognition and understanding. Furthermore, the sparse coding and SURF is used to deal with redundant image features, which is beneficial to run in real-time on desktop platforms for reducing computational complexity.

The contributions of this chapter are as follows: (1) A fusion framework based on D-S theory is introduced that utilizes data from RGB and depth sensors to realize gesture intention understanding, solving the new scientific problem of intent recognition using traditional methods. The recognition rates of fusion method were higher than other comparison methods and the condition when each sensor was used individually. (2) A salient high-level feature is constructed. The SURF algorithm is used to extract the features of the original image, and then perform the Sparse Coding, which is beneficial to improve intention recognition rate and run in real-time on desktop platforms. (3) The preliminary application experiments are also carried out in the developing emotional social robot system and the results indicate that the proposal can be applied to human-robot interaction.

The chapter is structured as follows. In Sect. 8.2, the details of gesture intention understanding are introduced, and experiments as well as analysis are presented in Sect. 8.3. The last section offers some conclusions.

8.2 Foreground Segmentation and Feature Extraction

Before performing the image feature extraction, it is necessary to extract the foreground containing human gesture information. For the depth foreground segmentation, we implement segmentation by detecting depth data components. As for the RGB foreground segmentation, we implement it by an iterative threshold method.

8.2.1 Foreground Segmentation Based on Depth and RGB Images

The Kinect can capture the depth data with user indexes [8]. The depth data with user index is made up of 16 binary digits, the upper 13 bits indicate the distance between the user and the device, and the lower 3 bits are the user index. The lower three bits are from 000 to 111, which respectively represent the background and user 1 to user 6. So we can easily segment the foreground image by merely retrieving the user index of the depth data.

As far as RGB image segmentation is concerned, we adopt the iterative threshold segmentation method to extract the gesture foreground [34]. It could obtain a relatively accurate segmentation with less computational overhead using only a single parameter. The segmentation algorithm works well when the histogram of the image is bimodal, and the trough is deeper and wider. In the iterative threshold segmentation, Z_{max} and Z_{min} are the maximum and the minimum gray value of the image, then the initial threshold T_0 is the average of Z_{max} and Z_{min}.

The convergence condition is that the difference of the threshold after the iteration is less than a predefined threshold, which determines the accuracy of the threshold convergence.

8.2.2 Speeded-Up Robust Features Based Gesture Feature Extraction

In gesture feature extraction, we adopt the Speeded-up Robust Features (SURF) to detect the local features of images, to achieve the invariance for rotation, scale, lightness as well as the stability for perspective transformation, affine change and noise. The SURF algorithm can detect the features with high speed and a low computing overhead because of the use of integral images method [14]. It uses a Hessian matrix to determine whether the points in the image are extreme points by searching for images in all scale spaces, which are potential points of interest that are not sensitive to scale and selection. For each point $f(x, y)$ of the image, the discriminant of the Hessian matrix is:

$$Det(H) = L_{xx} * L_{yy} - (L_{xy})^2 \tag{8.1}$$

where $L_{xx}(x, \delta) = \frac{\partial^2 g(\delta)}{\partial x^2} * I(x, y)$, which is the convolution of second-order Gaussian for integral image, and representing the Gaussian scale space of the image. δ denotes the scale parameter of the point (x, y).

In order to improve the calculation speed, the SURF algorithm uses a box filter instead of the Gaussian filter L [14]

$$Det(H) = D_{xx} D_{yy} - (w D_{xy})^2 \tag{8.2}$$

where $D_{(\cdot)}$ is the approximate solution corresponding to $L_{(\cdot)}$, and w is generally taken 0.9 according to [14], the role of which is to balance the error caused by the use of the box filter. Different scale of images can be obtained by using different size box filter templates, which could help in search for the extreme point of the speckle response.

The algorithm uses a non-maximum suppression algorithm to determine the initial feature points, searching for the maximum value in the neighbourhood and surrounding scale space around each pixel, and filtering out weaker points and error-positioned points to screen for the final stable feature point. And the main direction of the feature point is determined by the maximum value of the Harr wavelet feature in the circular neighborhood of the feature point.

For each point to be selected, we calculate the Haar wavelet features of 16 sub-regions in the square region around the feature points, and then get the following 4 values in each sub-region. The length of eigenvector for each candidate point is $16 \times 4 = 64$.

$$\left[\sum dx, \sum dy, \sum |dx|, \sum |dy|\right] \tag{8.3}$$

8.3 Encoding Speeded-Up Robust Features: Sparse Coding

The sparse coding (SC) algorithm is used to remove the redundant features and form more significant gesture image features representation, which is benefit for the image classification and recognition [35].

Based on the method of SURF algorithm, the training sample data set X is obtained. Assuming the basis vectors are φ_i, and the over-complete dictionary is $\varphi = \{\varphi_i\}_{i=1}^{n} \in \mathbb{R}^{d\times n}$, then the X can be represented as $X = \sum_{i=1}^{k} \alpha_i \varphi_i$, in which the α is the sparse coefficient.

For an over-complete base, the coefficient α is typically under-determined by the input vector X. Therefore, the "sparsity" standard is introduced to solve the problem. The purpose of sparse coding algorithm is to minimize the following objective function:

$$a\hat{} = \min_{\alpha,\varphi} \sum_{j=1}^{m} \left\| x^{(j)} - \sum_{i=1}^{k} \alpha_i^{(j)} \varphi_i \right\|^2 + \lambda \sum_{i=1}^{k} S(\alpha_i^{(j)}) \tag{8.4}$$

$$\text{subject to} \quad \|\varphi\|^2 \le C, \forall i = 1, ..., k$$

where λ is a regularization parameter which balances the influence of the reconstruction term and the sparsity term, m is the number of input vectors, k is the dimension of sparse coefficient, $S(.)$ is a sparse cost function. In practice, the general choice of the $S(.)$ is the $L1$ norm cost function $S(\alpha_i) = |\alpha_i|_1$ and the logarithmic cost function $S(\alpha_i) = \log(1 + \alpha_i^2)$.

SC has a training phase and a coding phase. Firstly, randomly initialize a visual dictionary φ and fix it, adjust α_i to minimize the objective function. Then the coefficient vector α is fixed, and we optimize the dictionary φ. Finally, we iterate the former two steps until convergence, and obtain the visual dictionary φ and the sparse code α of the training sample X. In the coding phase, for each image represented as a descriptor set X, the SC codes are obtained by optimizing (8.5) with respect to α only.

8.4 Multi-class Linear Support Vector Machines

Suppose that α is the result of sparse coding for feature descriptor X using Eq. (8.5). We construct the following pooling function as

$$\begin{aligned} z &= \mathscr{F}(\alpha) \\ z_j &= \max\left\{\left|\alpha_{1j}\right|, \left|\alpha_{2j}\right|, ..., \left|\alpha_{Mj}\right|\right\} \end{aligned} \tag{8.5}$$

where the pooling function $\mathscr{F}$ is defined on each column of α. Each column of α corresponds to the responses of all the local descriptors to one specific item in dictionary ϕ. Therefore, different pooling functions construct different image statistics. We define the pooling function $\mathscr{F}$ as a *max* pooling function on the absolute sparse codes. z_j is the jth element of z, α_{ij} is the matrix element at ith row and jth column of α, and M is the number of local descriptors present in the region.

Referring to the method of [23], we use multi-class linear SVM to realize the learning and classification of the features. According to the linear kernel, the decision function of the SVM can be constructed as follows:

$$f(z) = \sum_{i=1}^{n} \alpha_i k(z, z_i) + b = \left(\sum_{i=1}^{n} \alpha_i z_i \right)^T z + b = w^T z + b \tag{8.6}$$

The SC of the image features extracted by SURF generates sparse representations at different scales on the spatial pyramid. We construct a *max* pool function on different scales of the spatial pyramid by (11.10), and a pooling feature z is generated. A simple linear SPM kernel is constructed as follows:

$$k(z_i, z_j) = {z_i}^T z_j = \sum_{l=0}^{2} \sum_{s=1}^{2^l} \sum_{t=1}^{2^l} < z_i^l(s,t), z_j^l(s,t) > \tag{8.7}$$

where $< z_i, z_j >= {z_i}^T z_j$, and $z_i^l(s,t)$ is the max pooling statistics of sparse representations in the (s,t)th segment of image scale level l.

Assuming that the training set is $\{(z_i, y_i)\}_{i=1}^{n}$, $y_i \in \{1, ..., L\}$, L linear SVMs are trained using one-against-all strategy. The optimized objective of each SVM is as follows:

$$\min_{w_c} \left\{ J(w_c) = \|w_c\|^2 + C \sum_{i=1}^{n} \ell(w_c; y_i^c, z_i) \right\} \tag{8.8}$$

where y_i^c, z_i is the category label and $\ell(w_c; y_i^c, z_i)$ is the hinge loss function, which is defined as follows:

$$\ell(w_c; y_i^c, z_i) = \left[\max(0, w_c^T z \cdot y_i^c - 1)\right]^2 \tag{8.9}$$

8.5 Dempster-Shafer Evidence Theory for Decision-Level Fusion

Bayesian statistical inference and Dempster-Shafer evidence inference are two representative multi-sensor data fusion methods. The D-S theory is weaker than Bayesian statistical inference in terms of the existing restrictions because it does not require prior and conditional probabilities. So that we select the D-S theory method for decision fusion in image target recognition [36].

In the D-S evidence theory, assuming Δ is a hypothesis in the recognition framework Θ, the combination rules for two Basic Probability Assignment (BPA) $m(i)$ and $m(j)$ are as follows:

$$\begin{aligned} m_{i,j}(\Delta) &= \tfrac{1}{1-K} \sum_{\Delta_1 \cap \Delta_2 = \Delta} m_i(\Delta_1) m_j(\Delta_2) \\ K &= \sum_{\Delta_1 \cap \Delta_2 = \emptyset} m_i(\Delta_1) m_j(\Delta_2) \end{aligned} \tag{8.10}$$

where K is the normalization factor, while $1 - K$ reflects the degree of evidence conflict.

According to [37], for the frame of discernment $\Theta = [\Theta_1, \Theta_2, ..., \Theta_n]$, where n is the number of gesture classes, we define $\ell(y)$ as the hinge loss function in the linear SVM optimization object. The BPA over Θ can be constructed as follows:

$$\begin{aligned} m(\Theta_j \mid \tilde{y}_j) &= \beta \phi_j(\ell_j(y)) \\ m(\Theta \mid \tilde{y}_j) &= 1 - \beta \phi_j(\ell_j(y)) \\ m(D \mid \tilde{y}_j) &= 0, \forall D \in 2^{\Theta} \setminus \{\Theta, \Theta_j\} \end{aligned} \tag{8.11}$$

where β is a control parameter and $0 < \beta < 1$, $\tilde{y}_j$ is each class-specific representation. If y is close to $\tilde{y}_j$, it is most likely that H_j is true. If $\ell(y)$ is large, the class of $\tilde{y}_j$ will provide little or no information about the class of y. ϕ_j is a decreasing function that satisfies the following conditions:

$$\begin{aligned} \phi_j(0) &= 1 \\ \lim_{\ell(y) \to \infty} \phi_j(\ell(y)) &= 0 \end{aligned} \tag{8.12}$$

According to [37], we determine ϕ_j as follows:

$$\phi_j(\ell(y)) = e^{-\ell(y)^2} \tag{8.13}$$

According to the Dempster's rule of combination, the global BPA of combining n BPAs, is shown as follows:

$$\begin{aligned} & m_c(\Theta_j) = \tfrac{1}{K_0} \{\beta \phi_j(\ell_j(y))\} \cdot \prod_{i \neq j} \{1 - \beta \phi_i(\ell_i(y))\}, \\ & i \in \{1, ..., n\} \\ & m_c(\Theta) = \tfrac{1}{K_0} \prod_{j=1}^{n} \{1 - \beta \phi_j(\ell_j(y))\} \\ & K_0 = \sum_{j=1}^{n} \prod_{p \neq j} \{1 - \beta \phi_p(\ell_p(y))\} + (1-n) \prod_{j=1}^{n} \{1 - \beta \phi_j(\ell_j(y))\} \end{aligned} \tag{8.14}$$

After calculation, the global BPAs of depth and RGB data are obtained, which can be recorded as m_c1 and m_c2. The combined BPA from m_c1 and m_c2 is then obtained via (8.11). Then the confidence function $Bel(\Theta_j)$ is obtained to identify the category of the gesture, which yields the maximum value of the $Bel(\Theta_j)$.

8.6 Experiments

In this part, the data settings and experimental environment settings are explained. The simulation experiment is carried out and the experimental results are analyzed. To verify the effectiveness of the proposal, experiments on two RGB-D datasets (CGD2011 and CAD-60) are conducted.

8.6.1 Experimental Setting

The Cornell Activity Datasets (CAD-60) [38] and ChaLearn Gesture Dataset (CGD 2011) [39] are used for the experiment. The Cornell Activity Datasets (CAD-60) is an RGB-D database collected by members of the Robot Learning Lab at Cornell University. It uses Microsoft Kinect to record 12 actions of 4 people (2 men, 2 women) in 5 different environments. The resolution of the RGB-D image is 320*240. We select seven different gestures in the living room environment of the 4 people, as shown in Fig. 8.1. Based on the considerations of data volume, we randomly selected 200 RGB-D images in each type of gesture, that is, a total of 5600 images for this database. The CGD 2011 is a human gesture database in video format recorded by Kinect, which includes both depth and RGB information [26]. We also selected representative seven types of hand gestures (5600 images) for intention understanding as shown in Fig. 8.2. The size of each frame is 320*240. Original images are labelled 1 to 7 for these seven kinds of gestures, representing seven kinds of gesture intentions, respectively.

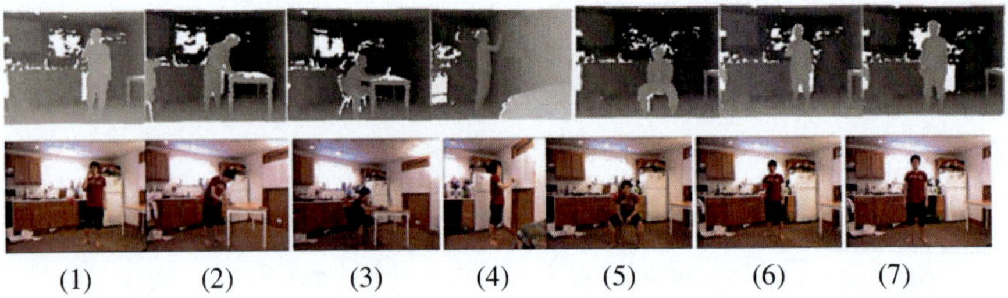

Fig. 8.1 Images of 7 gestures intentions (CAD-60). (1) Talking on the phone (2) Cooking (chopping) (3) Working on computer (4) Writing on whiteboard (5) Talking on couch (6) Drinking water (7) Standing still

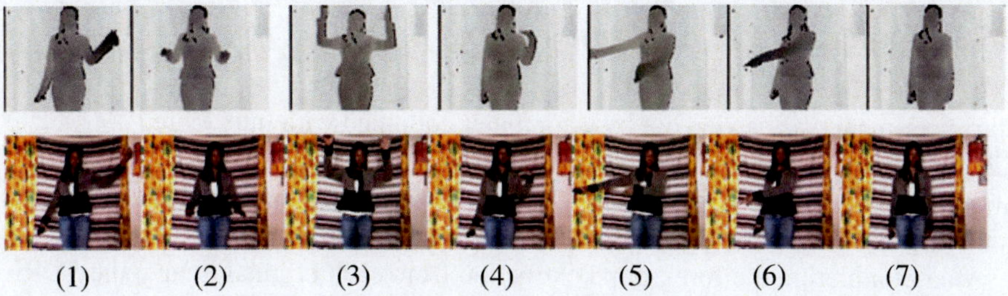

Fig. 8.2 Images of 7 gestures intentions (CGD2011)

To verify the effectiveness of the proposed decision level fusion framework, we compared the recognition performance of our framework with the performance of each single-modal image sensor. Overall three experiments are executed and evaluated. The first one uses information of the RGB images to recognize seven gesture intentions. The second one considers information coming from the depth images, and the third one uses both the two kinds of information. Meanwhile, we use different classifiers for comparison, and we also compared the effects of sparse coding on the experimental results. In the first comparative experiment, we eliminated the step of sparse coding, and directly input the extracted SURF features into the SVM for classification and identification, finally use D-S evidence theory for data fusion, which is to verify the effect of sparse coding on image features processing. In the second comparative experiment, we use BP neural network (BPNN) instead of SVM to achieve feature classification, which is to verify the classification of SVM. Furthermore, referring to [37], we use the sparse representation classifier (SRC) to classify SURF features instead of SVM, and finally use D-S evidence theory for decision-level fusion. In addition, we use the bagging method as a comparison to verify the effectiveness of the D-S theory in the fusion framework. The Bootstrap Aggregating algorithm is used to combine multiple SVM classifiers to improve the classification effect. We also used a random forest approach to mitigate the effects of strong classifiers (SVMs) on ensemble process. The Bootstraping method is used to randomly extract 80% of the samples from the training data as the training set and the remaining 20% as the verification set. This process will be performed in 10 rounds (some samples may be extracted multiple times), and 10 training models are obtained. The finally classification results of the verification set and the testing set are obtained by voting method for each models. To verify the performance of the experimental results, the cross validation test (CV test) of 10-fold and leave-one-subject-out (LOSO) were considered to remove any bias. In the 10-fold CV test, we divide the samples into 10 equal parts, and select 9 parts of them to train the model, and the remaining 1 part to test the modal. The above process will iterate for 10 times. In the LOSO CV test, we used the data from 3 of the 4 people in the sample to train the model, and leaving 1 person's data for testing, and then iterating the process 4 times.

8.6.2 Experimental Environment and Setup

The experiment was carried out by a computer equipped with Intel Core i7-7700HQ CPU processor and 2.8 GHz system clock as well as a RAM of 8.00 GB. The software environment is Matlab R2017a. The Datasets of CGD2011 and CAD-60 were used in the experiment to verify the effectiveness of the proposal.

After completing the foreground extraction from each original image, the SURF feature extraction is performed on the foreground images, and the training sample data set X is obtained, which is a 3450-dimensional feature descriptor for each image. It prescribed that 50 feature points are extracted for each image, and the extracted feature information includes landmark position, the scale of the detected landmark, Laplacian of the landmark neighborhood, orientation in radians and 64-dimensional feature descriptor, which is 69-dimensional data. We arrange these data to form a $69 * 50 = 3450$ dimensional feature vector.

Next, the SURF feature descriptors are subjected to sparse coding, and α and ϕ_i are iteratively trained using Eq. (11.9) to obtain codebook ϕ_i. Then, all SURF feature descriptors are coded using the codebook ϕ_i to solve the coding coefficient α, and it has 5376-dimensional features in the experiment. The regularization parameter λ in the SC is set to 0.15, which is proved to be better after plenty of experiments. Finally, the gesture classification is realized by linear SVM, and we set the coefficient for regularization of SVM to 0.2. The decision-level data fusion of two kinds of image recognition is realized by D-S evidence theory. As suggested in [37], we set $\beta = 0.95$ for the BPA in (10.19).

In the comparative experiment, we construct a three-layer BP neural network, in which the number of hidden layer nodes is 30. And we set the iterations times to 5000, and the learning rate is 0.05, the modal will stop iteration when the error is less than 0.1.

8.6.3 Experimental Results and Analysis

The experimental results on CAD-60 are shown in Table 8.1. In the first case, the information of the RGB images and depth images are used to recognize seven kind of gesture intentions respectively, and the average recognition rate are 86.33 and 82.08%. However, the average recognition rate of fusion framework is 94.02%, which is obviously high up to 7.69 and 11.94% than the single-modal method, and it is higher than the latest recognition rate for this database [40]. We note that the recognition accuracy obtained by using the fusion method is better than the recognition using depth and RGB data alone, which reflects the importance of the fusion framework for the accuracy of recognition results. Figure 8.3 shows the confusion matrices of the recognition result.

In the second case, the 3,450-dimensional SURF image features are directly input into the SVM for classification and recognition, and the final recognition rate is

Table 8.1 Intention understanding results of 10-fold cross test (CAD-60)

Index	1	2	3	4	5	6	7	8	9	10	Avg±std
Proposal (%)	96.68	94.15	94.23	89.99	94.25	92.07	92.65	95.98	92.59	97.62	94.02±2.31
RGB (%)	86.95	86.18	89.03	81.35	91.70	85.14	86.58	86.21	83.95	86.19	86.33±2.62
Depth (%)	84.78	80.15	84.70	81.09	84.46	79.07	77.94	85.23	80.12	83.24	82.08±2.57
without SC (%)	72.29	68.95	75.47	83.24	79.78	76.71	61.90	80.00	73.11	69.00	74.04±6.37
BPNN (%)	83.69	72.64	81.61	75.26	86.19	85.21	84.32	80.51	79.25	83.98	81.27±4.21
SRC [37] (%)	83.36	82.23	82.52	85.61	80.26	81.58	86.98	85.32	83.20	85.25	83.63±2.10
Bagging-SVM (%)	89.51	83.42	88.45	85.08	93.24	87.19	80.26	95.81	90.01	91.23	88.42±4.40
Random Forest (%)	95.45	89.98	93.68	88.24	93.01	90.29	97.01	91.93	96.72	94.26	93.06±2.80
MF [40] (%)	/	/	/	/	/	/	/	/	/	/	89.00

	(1)	(2)	(3)	(4)	(5)	(6)	(7)
(1)	0.83	0.02	0	0	0	0.08	0.07
(2)	0.03	0.86	0.09	0.02	0	0	0
(3)	0.04	0.11	0.75	0	0.07	0.03	0
(4)	0.09	0.06	0	0.76	0	0.04	0.05
(5)	0.02	0	0.07	0	0.89	0.02	0
(6)	0.05	0	0	0	0	0.84	0.11
(7)	0.07	0	0	0.02	0	0.08	0.82

(a) Depth image

	(1)	(2)	(3)	(4)	(5)	(6)	(7)
(1)	0.82	0	0	0.02	0	0.1	0.06
(2)	0	0.96	0.01	0.03	0	0	0
(3)	0	0.04	0.89	0	0.07	0	0
(4)	0	0.08	0	0.88	0	0.04	0
(5)	0	0	0.05	0	0.95	0	0
(6)	0.1	0	0	0.01	0	0.77	0.12
(7)	0.12	0	0	0.02	0	0.13	0.73

(b) RGB image

	(1)	(2)	(3)	(4)	(5)	(6)	(7)
(1)	0.92	0.01	0	0	0	0.04	0.03
(2)	0	0.97	0.01	0.02	0	0	0
(3)	0	0.07	0.93	0	0.04	0	0
(4)	0.02	0.02	0	0.94	0	0	0.02
(5)	0	0	0.01	0	0.98	0.01	0
(6)	0.02	0	0	0	0	0.92	0.06
(7)	0.04	0	0	0	0	0.05	0.91

(c) Fusion result

Fig. 8.3 Confusion matrix of the recognition result (CAD-60) (1) Talking on the phone (2) Cooking (chopping) (3) Working on computer (4) Writing on whiteboard (5) Talking on couch (6) Drinking water (7) Standing still

74.04% with a deviation of 6.37%, which indicates that redundant image features have a bad influence on the classification effect of SVM.

In the third case, in order to verify the impact of SVM model on the classification effect, we use BPNN instead of SVM to achieve feature classification. And the final recognition rate is 81.27% with a deviation of 4.21%, which indicates that the SVC can achieve better results than BPNN in image classification with a certain amount of data.

In the fourth case, in order to verify the impact of the SVM model on the classification effect, we use the sparse representation classifier (SRC) to classify SURF features instead of SVM referring to [37], and finally D-S evidence theory was used for decision-level fusion. The final recognition rate is 83.63%, which is 2.10% lower than the SVM method and proves the stability of the SVM classifier.

In the fifth case, we use the bagging method for comparison experiments to verify the effect of the D-S theory. The final recognition rate is 88.42% with SVM for classifier, and 93.06% with the Random Forest method, which may result by the over-fitting caused by small data volume with complex model structure. In addition, we investigated other research results on the database [40]. Modality-based fusion (MF) method combines the skeleton features in the database, and constructs local

Table 8.2 Intention understanding results of 10-fold cross test (CGD2011)

Index	1	2	3	4	5	6	7	8	9	10	Avg±std
Proposal (%)	93.06	93.36	91.53	92.56	92.26	91.50	89.96	91.56	92.48	94.53	92.28±1.25
RGB (%)	93.32	85.56	85.33	86.52	92.65	91.41	93.01	94.65	94.21	83.64	90.03±4.25
Depth (%)	93.21	90.78	80.66	90.31	80.26	90.35	92.22	90.33	85.65	78.33	87.21±5.53
without SC (%)	88.63	80.52	72.95	80.64	82.65	78.22	80.54	80.00	71.22	67.23	78.26±6.19
BPNN (%)	85.62	90.71	69.36	79.43	85.69	89.28	86.61	84.29	72.15	81.32	82.45±6.66
SRC [37] (%)	90.40	90.43	87.21	85.70	86.35	86.63	81.64	90.69	85.20	87.25	87.14±2.81
Bagging-SVM (%)	92.45	87.38	90.12	90.76	88.62	91.04	94.01	89.65	93.82	88.86	90.67±2.10
Random Forest (%)	94.79	90.31	93.09	89.64	88.42	93.87	93.65	87.01	95.90	93.98	92.07±2.84

	(1)	(2)	(3)	(4)	(5)	(6)	(7)
(1)	0.89	0.01	0.02	0.08	0	0	0
(2)	0	0.91	0.05	0.02	0.02	0	0
(3)	0	0.1	0.84	0.04	0.02	0	0
(4)	0.09	0.03	0.02	0.86	0	0	0
(5)	0	0	0.01	0	0.93	0.06	0
(6)	0.01	0	0.02	0	0.05	0.92	0
(7)	0	0.05	0	0.02	0	0.03	0.90

(a) RGB image

	(1)	(2)	(3)	(4)	(5)	(6)	(7)
(1)	0.86	0.02	0.03	0.09	0	0	0
(2)	0.01	0.89	0.06	0.03	0	0.01	0
(3)	0.01	0.08	0.87	0.04	0	0	0
(4)	0.09	0.03	0	0.87	0.01	0	0
(5)	0	0.02	0	0	0.90	0.08	0
(6)	0	0.01	0	0	0.05	0.94	0
(7)	0.01	0.05	0	0.03	0	0.01	0.91

(b) Depth image

	(1)	(2)	(3)	(4)	(5)	(6)	(7)
(1)	0.90	0.02	0.02	0.06	0	0	0
(2)	0	0.94	0.06	0	0	0	0
(3)	0.02	0.06	0.89	0.03	0	0	0
(4)	0.06	0.04	0.02	0.88	0	0	0
(5)	0	0	0	0	0.95	0.05	0
(6)	0	0	0	0	0.04	0.96	0
(7)	0	0.05	0	0.01	0	0	0.94

(c) Fusion result

Fig. 8.4 Confusion matrix of the recognition result (CGD2011)

Table 8.3 CV test of leave-one-subject-out (LOSO) for the two databases

Proposal	CAD-60	CGD2011
10-fold CV (%)	94.02	92.28
LOSO CV (%)	92.62	91.24

descriptors respectively to achieve action classification, which finally produced the accuracy of 89%.

The experimental results on CGD2011 are shown in Table 8.2. Here we completed the same comparative experiment as for the CAD-60 database, and average accuracies of the five method are 92.28% of the proposed model, 90.03% of RGB images, and 87.21% of depth images. When we discard the sparse coding steps and directly use SVM to classify SURF features, the result is 78.26%. And the recognition result are 82.45 and 87.14% when using the BPNN and SRC. When we use the ensemble techniques, the final results are 90.67 and 92.07%. The comparison results are shown in Fig. 8.4.

In order to verify the validity of the experimental result, we made a CV test of leave-one-subject-out (LOSO) for the two databases, and the samples of one person were used as the test data, and samples of the other three people were used as the training data, and the final result is shown in Table 8.3.

8.7 Summary

In this chapter, a fusion framework is introduced that utilizes data from RGB and depth sensors to realize gesture recognition. Sparse coding algorithm based on SURF feature is used for gesture feature extraction, and linear SVM is used for feature classification, and finally recognition results from two type of sensors are achieving decision-level fusion by D-S evidence theory. To verify the effectiveness of the proposal, experiments on two RGB-D datasets (CGD2011 and CAD-60) are conducted, the experimental results show that the recognition rates are improved by using these two different modality sensors together compared to the situations when each sensor is used individually.

Inspired by the preliminary application, the proposed intention understanding model can be applied to the development of an emotional social robot system. The model's identification of human gesture intention can help emotional robot to provide more proactive and high quality services.

References

1. L.F. Chen, M.T. Zhou, M. Wu, J.H. She, Z.T. Liu, F.Y. Dong, Three-layer weighted fuzzy support vector regression for emotional intention understanding in human-robot interaction. IEEE Trans. Fuzzy Syst. **26**(5), 2524–2538 (2018)
2. I. Andrey, Real-time human activity recognition from accelerometer data using convolutional neural networks. Appl. Soft Comput. **62**, 915–922 (2019)
3. C.H. Hsu, S. Wang, Y. Yuan, Guest editorial special issue on hybrid intelligence for internet of vehicles. IEEE Syst. J. **11**(3), 1225–1227 (2017)
4. J. Yang, Y. Wang, Z. Lv et al., Interaction with three-dimensional gesture and character input in virtual reality: recognizing gestures in different directions and improving user input. IEEE Consum. Electron. Mag. **7**(2), 64–72 (2018)
5. A. Kleinsmith, N. Bianchiberthouze, Affective body expression perception and recognition: a survey. IEEE Trans. Affect. Comput. **4**(1), 15–33 (2013)
6. L.F. Chen, M.T. Zhou, W. Su, M. Wu, J.H. She, K. Hirota, Softmax regression based deep sparse autoencoder network for facial emotion recognition in human-robot interaction. Inf. Sci. **428**, 49–61 (2018)
7. S.C. Neoh, L. Zhang, K. Mistry, M.A. Hossain, Intelligent facial emotion recognition using a layered encoding cascade optimization model. Appl. Soft Comput. **34**, 72–93 (2015)
8. B.I. Ahmad, J.K. Murphy, P.M. Langdon et al., Intent inference for hand pointing gesture-based interactions in vehicles. IEEE Trans. Cybern. **46**(4), 878–889 (2015)
9. A. Mohanty, R.R. Sahay, Rasabodha: understanding Indian classical dance by recognizing emotions using deep learning. Pattern Recogn. **79**, 97–113 (2018)
10. Y. Feng, L.F. Chen, W.J. Su, K. Hirota, Gesture intention understanding based on depth and RGB data, in *Proceedings of the 37th Chinese Control Conference* (2018), pp. 984–987
11. J. Han, E.J. Pauwels, P.M.D. Zeeuw et al., Employing a RGB-D sensor for real-time tracking of humans across multiple re-entries in a smart environment. IEEE Trans. Consum. Electron. **58**(2), 255–263 (2012)
12. C.J. Su, C.Y. Chiang, J.Y. Huang, Kinect-enabled home-based rehabilitation system using Dynamic Time Warping and fuzzy logic. Appl. Soft Comput. **22**(5), 652–666 (2014)
13. D.G. Lowe, Distinctive image features from scale-invariant keypoints. Int. J. Comput. Vis. **60**(2), 91–110 (2004)

14. H. Bay, A. Ess, T. Tuytelaars, L.V. Gool, SURF: speeded up robust features, in *Proceedings of Computer Vision and Image Understanding (CVIU)* (2008), pp. 346–359
15. A. Jaszkiewicz, T. Lust, ND-Tree-based update: a fast algorithm for the dynamic non-dominance problem. IEEE Trans. Evol. Comput. **22**(5), 778–791 (2018)
16. E. Phaisangittisagul, S. Thainimit, W. Chen, Predictive high-level feature representation based on dictionary learning. Expert Syst. Appl. **69**, 101–109 (2017)
17. B. Li, F. Zhao, Z. Su, Example-based image colorization using locality consistent sparse representation. IEEE Trans. Image Process. **26**(11), 5188–5202 (2017)
18. B. Stefania, C. Alfonso, M.V. Peelen, View-invariant representation of hand postures in the human lateral occipitotemporal cortex. NeuroImage **181**, 446–452 (2018)
19. R.K. Lama, J. Gwak, J.S. Park et al., Diagnosis of alzheimer's disease based on structural MRI images using a regularized extreme learning machine and PCA features. J. Healthc. Eng. **1**, 1–11 (2017)
20. J. Wright, Y. Ma, J. Mairal, G. Sapiro, T.S. Huang, S. Yan, Sparse representation for computer vision and pattern recognition. Proc. IEEE **98**(6), 1031–1044 (2010)
21. B. Cheng, L. Jin, G. Li, General fusion method for infrared and visual images via latent low-rank representation and local non-subsampled shearlet transform. Infrared Phys. Technol. **92**, 68–77 (2018)
22. A. Helmi, M.W. Fakhr, A.F. Atiya, Multi-step ahead time series forecasting via sparse coding and dictionary based techniques. Appl. Soft Comput. **69**, 464–474 (2018)
23. J. Yang, K. Yu, Y. Gong et al., Linear spatial pyramid matching using sparse coding for image classification, in *Proceedings of IEEE Conference on Computer Vision and Pattern Recognition* (2009), pp. 1794–1801
24. Z. Liu, W. Zhang, S. Lin et al., Heterogeneous sensor data fusion by deep multimodal encoding. IEEE J. Sel. Top. Signal Process. **11**(3), 479–491 (2017)
25. Y. Zhang, B. Song, X. Du et al., Vehicle tracking using surveillance with multimodal data fusion. IEEE Trans. Intell. Transp. Syst. **19**(7), 2353–2361 (2018)
26. O. Katz, R. Talmon, Y.L. Lo et al., Alternating diffusion maps for multimodal data fusion. Inf. Fusion **45**, 346–360 (2018)
27. Y. Ma, Y. Hao, M. Chen, J. Chen, P. Liiu, Audio-visual emotion fusion (AVEF): a deep efficient weighted approach. Inf. Fusion **46**, 184–192 (2018)
28. M.L. Gavrilova, M. Monwar, Multimodal biometrics and intelligent image processing for security systems. Register 69–79 (2013)
29. J. Chaney, E.H. Owens, A.D. Peacock, An evidence based approach to determining residential occupancy and its role in demand response management. Energy Build **125**, 254–266 (2016)
30. R. Boukezzoula, D. Coquin, T.L. Nguyen et al., Multi-sensor information fusion: combination of fuzzy systems and evidence theory approaches in color recognition for the NAO humanoid robot. Robot. Auton. Syst. **100**, 302–316 (2018)
31. Q.F. Zhou, H. Zhou, Q.Q. Zhou, F. Yang, Structural damage detection based on posteriori probability support vector machine and Dempster-Shafer evidence theory. Appl. Soft Comput. **36**, 368–374 (2015)
32. C. Lu, S. Wang, X. Wang, A multi-source information fusion fault diagnosis for aviation hydraulic pump based on the new evidence similarity distance. Aerosp. Sci. Technol. **71**, 392–401 (2017)
33. L.F. Chen, M. Wu, M.T. Zhou, J.H. She, F.Y. Dong, K. Hirota, Information-driven multi-robot behavior adaptation to emotional intention in human-robot interaction. IEEE Trans. Cogn. Dev. Syst. **10**(3), 647–658 (2018)
34. D. Wang, H. Li, X. Wei et al., An efficient iterative thresholding method for image segmentation. J. Chem. Phys. **350**, 657–667 (2017)
35. X. Zhu, X. Li, S. Zhang et al., Robust joint graph sparse coding for unsupervised spectral feature selection. IEEE Trans. Neural Netw. Learn. Syst. **28**(6), 1263–1275 (2017)
36. L. F. Chen, M. Wu, M.T. Zhou, Z.T. Liu, J.H. She, K. Hirota, Dynamic emotion understanding in human-robot interaction based on two-layer fuzzy SVR-TS model. IEEE Trans. Systems. Man, Cybern. **50**(2), 490–501 (2020)

37. C. Chen, R. Jafari, N. Kehtarnavaz, Improving human action recognition using fusion of depth camera and inertial sensors. IEEE Trans. Hum.-Mach. Syst. **45**(1), 51–61 (2015)
38. J. Sung, C. Ponce, B. Selman et al., Unstructured human activity detection from RGBD images. Comput. Sci. **44**(8), 47–55 (2012)
39. I. Guyon, V. Athitsos, P. Jangyodsuk et al., The chaLearn gesture dataset (CGD 2011). Mach. Vis. Appl. **25**(8), 1929–1951 (2014)
40. B. Seddik, S. Gazzah, E.B.A. Najoua, Human-action recognition using a multi-layered fusion scheme of Kinect modalities. IET Comput. Vis. **11**(7), 530–540 (2017)

Chapter 9
Three-Layer Weighted Fuzzy Support Vector Regressions for Emotional Intention Understanding

A three-layer weighted fuzzy support vector regression (TLWFSVR) model is proposed for understanding human intention, and it is based on the emotion-identification information in human-robot interaction. TLWFSVR model consists of three layers, including adjusted weighted kernel fuzzy c-means (AWKFCM) for data clustering, fuzzy support vector regressions (FSVR) for information understanding, and weighted fusion for intention understanding. It aims to guarantee the quick convergence and satisfactory performance of the local FSVR via adjusting the weights of each feature in each cluster, in such a way that importance of different emotion-identification information is represented. Moreover, smooth human-oriented interaction can be obtained by endowing robot with human intention understanding capability.

9.1 Introduction

With the rapid growth of various robots presence in the human society, robots interact more and more with people. Social robots, including humanoid, bionic, and mechanical, become close parters of humans, so that human-robot interaction (HRI) has become inevitable [1–3]. Social robots promise widespread integration into human social and physical environment, and they have been placed in museums/exhibition expos, receptions, shopping malls, and other public areas [4–6], in such a way that HRI becomes increasingly socially situated and multi-faceted [7]. As these various applications have been shown, social robots have a promising future of not only attracting people by their novelty, but also being able to provide useful and reliable services in our daily life.

L. Chen et al., *Emotion Recognition and Understanding for Emotional Human-Robot Interaction Systems*, Studies in Computational Intelligence 926,
https://doi.org/10.1007/978-3-030-61577-2_9

To achieve smooth, natural, and rational HRI, robotic systems are developed in different fields with regard to adaptation [8], learning [9], autonomous behaviors [10], and robot control including parameter turning [11] and uncertainties compensation [12]. Many researches design adaptive strategies to adjust robots' behaviors for smooth communication [13, 14]. However, the interaction effect is less than satisfactory, and this is because robots lack of understanding about people's inner thoughts, e.g., intentions, that means, interaction only via social behaviors is not enough.

Human intention understanding is significant for social communication in our daily life, including image intention understanding [15], speech intention understanding [16], and gesture intention understanding [17]. Speech and gesture have great impact on intention understanding, but they cannot intuitively reflect people's inner thoughts. Emotion is considered as the driving force behind motivation on behavior, which is an important missing link in the intention-behavior gap [18], and emotion is more associated with people's inner thoughts in a psychological perspective. Therefore, emotion is supposed to be a key component to deeply reflect people's inner thoughts, and how to use individual emotion for intention understanding becomes necessary and feasible, which is also a new challenge.

As a powerful and promising learning machine, support vector machines (SVMs) are famous for solving problems characterized by small samples, nonlinearity, high dimensionality and local minima [19, 20]. With the introduction of ε-insensitive loss function, the SVM is extended to solve regression problems, called support vector regression (SVR) [21]. SVR has been researched during the past decade to deal with recognition, classification, or modeling problems in image processing [22] and pattern recognition [23]. It is being expanded to apply in the field of HRI.

SVR meets some challenges. One is the difficulty to guarantee the good local modeling performance, which means only one regression hyperplane cannot be guaranteed most training samples lie within it. Fuzzy support vector regression (FSVR) [24] has been proposed to enhance the adaptive and learning capability in complicated engineering situations. Referred to the idea of local learning [25], the use of FSVR is represented as fuzzy rules with the membership function in general [26, 27], and the number of rules is set to be identical to the number of SVRs. Each local SVR attempts to locally adjust the capacity to the properties of the training data subset.

Even in the local FSVR, there are still problems. The number of training data in each subset must be specified by users, if the subset is not well classified, the local SVR approach needs more computational time. Moreover, to prevent some data may lie outside the range of the subset, a large boundary bias may be produced. Consequently, the satisfactory performance is not guaranteed while different numbers of training data is considered in subsets. For considering such problems, FSVR with FCM algorithm is proposed, such as the two-layer fuzzy support vector regression (TLFSVR) in our previous study [28], and the training subsets is automatically clustered by FCM algorithm.

Nevertheless, as the age of big data is coming, especially huge complex information for deep level understanding, some problems may exist in the FSVR with FCM algorithm (e.g., TLFSVR in our previous study [29]). First, the importance of each cluster in FCM is the same, and this will result in the satisfactory performance of the local SVRs fusion cannot be guaranteed. Second, the weights of each feature in each cluster is the same, hence, modeling of each local SVR will be not satisfied. Third, distance between input data and clustering center is defined as Euclidean distance in FCM, which is largely limited to handle the data with similar scale. To overcome the above problems, three-layer weighted fuzzy support vector regression (TLWFSVR) is proposed. Instead of FCM in the TLFSVR [30], the proposed adjusted weighted kernel fuzzy c-means (AWKFCM) is used to data clustering in the new TLWFSVR. In the AWKFCM, adjusted weights are used to represent the importance of each cluster, which improves the effectiveness of data clustering and SVRs learning. Moreover, the kernel distance enables data clustering to operate in a high-dimensional, implicit feature space. It is no need to compute the coordinates of the data, but rather by simply calculating the inner products between all pairs of data in the feature space.

In the proposed TLWFSVR, age, gender, and nationality are additional information for dealing with the problem of various people coexistence. Three layers include adjusted weighted kernel fuzzy c-means (AWKFCM) for data clustering, fuzzy support vector regressions (FSVR) for information understanding, and weighted fusion for intention understanding. First, data clustering by using the proposed AWKFCM. Then, the training data is split into multiple subsets. Second, in view of the local learning [31], the traditional one SVR is extended to multiple local SVRs. For example, people with different ages/genders/nationalities may cause problems such as aging, gender hobby, religion, and so on. Different local SVRs reflect different ages/genders/nationalities, which is better than only one regression model covering all the situations. Finally, fuzzy inference based on identification information is used to generate the intention.

The proposal endows robot with human-oriented capabilities to deeply understand people's intention, and robot communicates with the user or guess what he/she expects. Human-oriented communication means that robot is able to consider human information, including identification information (e.g., age, gender, and nationality) and deep-level information (e.g., intention and emotion). Moreover, our proposal effectively classifies the information by considering the different impacts of each one, which facilitates the intention understanding, so that the communication may run smoothly.

Simulation and preliminary application experiments are developed to show the effectiveness of the proposed TLWFSVR based intention understanding model. Simulation experiments are designed to verify the accuracy of the proposal, where they are built by Matlab in a PC employed with a dual core processer (2.0G Hz and 2.6 GHz), memory (8 GB), and Windows 8 system. In addition, to confirm the validity and practicability of the proposal, preliminary application experiments are conducted in the developing emotional social robot system (ESRS), where a scenario of "drinking at bar" is performed by twelve volunteers and two mobile robots.

9.2 Support Vector Regression

SVM employs structural risk minimization (SRM) principle to achieve better generalization ability, so that it provides higher performance than traditional empirical risk minimization (ERM) based learning machines, neural networks [29]. There are two main categories for SVM, which are support vector classification (SVC) and support vector regression (SVR). While SVC is mostly used to perform classification by determining the maximum margin separation hyperplane between two classes, SVR tries the inverse, i.e., to find the optimal regression hyperplane so that most training samples lie within an ε-margin around this hyperplane. Suppose training data set is a set of l pairs of vectors and the set is described as $D = \{(X_i, y_i)\,|X_i \in R^n, y_i \in R, i = 1, 2, \ldots, l\}$. Considering the simplest case of linear SVR, the basic function is given as

$$f(X) = \langle W, X \rangle + b \tag{9.1}$$

where the function $f(X)$ involves taking the dot product of the input vector X with a model parameter W and adding the scalar bias b.

The aim of ε-SVR is to find a function $f(X)$ that has at most ε deviation from the actually obtained targets y_i for all the training data, and at the same time as flat as possible by minimizing the norm of W. The problem can be described as convex optimization problem and written as

$$\begin{array}{ll} \min & \frac{1}{2}\,\|W\|^2 \\ \text{subject to} & y_i - \langle W, X_i \rangle - b \leqslant \varepsilon \\ & \langle W, X_i \rangle + b - y_i \leqslant \varepsilon. \end{array} \tag{9.2}$$

Sometimes constrains may not always be satisfied, that means not all pairs (X_i, y_i) fit with ε precision. Slack variables ξ_i and ξ_i^* are introduced to cope with the otherwise infeasible constrains, and (9.2) becomes

$$\begin{array}{ll} \min\limits_{W,b,\xi_i,\xi_i^*} & \frac{1}{2}\,\|W\|^2 + \gamma \sum\limits_{i=1}^{l} \left(\xi_i + \xi_i^*\right) \\ \text{subject to} & y_i - \langle W, X_i \rangle - b \leqslant \varepsilon + \xi_i \\ & \langle W, X_i \rangle + b - y_i \leqslant \varepsilon + \xi_i^* \\ & \xi_i, \xi_i^* \geqslant 0, i = 1, 2, \ldots, l \end{array} \tag{9.3}$$

where γ is a scalar regularization parameter that penalizes the amount of slack used (smaller values of γ yield more outlier rejection).

Retrieving the optimal hyperplane in (9.3) is a QP problem, which can be solved by constructing a Lagrangian and transformed into the dual as (9.4), where α_i and α_i^* are the Lagrange multipliers. All the dot product in (9.4) can be replaced by kernel function and the regression function is reconstructed as

$$\begin{aligned}\max_{\alpha,\alpha^*} \quad & -\tfrac{1}{2}\sum_{i,j=1}^{l}\left(\alpha_i-\alpha_i^*\right)\left(\alpha_j-\alpha_j^*\right)\langle X_i, X_j\rangle \\ & -\varepsilon\sum_{i=1}^{l}\left(\alpha_i+\alpha_i^*\right)+\sum_{j=1}^{l} y_i\left(\alpha_i-\alpha_i^*\right) \\ \text{subject to} & \sum_{i=1}^{l}\left(\alpha_i-\alpha_i^*\right)=0 \quad and \quad \alpha_i,\alpha_i^*\in[0,\gamma]\end{aligned} \tag{9.4}$$

$$f(X)=\sum_{i=1}^{l}\left(\alpha_i-\alpha_i^*\right)K(X_i,X)+b \tag{9.5}$$

where the kernel function is the Gaussian kernel,

$$K(x_i,x_j)=\exp\left(-\frac{(x_i-x_j)^2}{2\sigma^2}\right). \tag{9.6}$$

9.3 Three-Layer Fuzzy Support Vector Regression

Three-layer includes data clustering, information understanding and intention understanding layer, as shown in Fig. 9.1.

The training data sets are split into several subsets by using the adjusted weighted kernel fuzzy c-means (AWKFCM) algorithm. In Kernel fuzzy c-means (KFCM) [30, 31], considering nonlinear mapping function $\phi : \mathrm{x} \to \phi(\mathrm{x}) \in R^{D_K}$, where D_K is the dimensionality of the transformed feature vector x. With kernel clustering, explicitly transform x is not necessary, and we simply need to represent the dot product $\kappa(\mathrm{x}, \mathrm{x}) = \phi(\mathrm{x}) \cdot \phi(\mathrm{x})$, with the radial basis function (RBF)

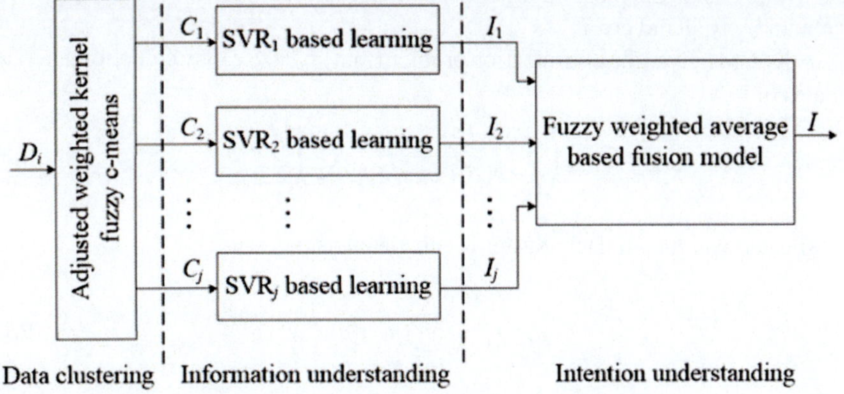

Fig. 9.1 Three-layer weighted fuzzy support vector regression (TLWFSVR)

$\kappa(\mathrm{x}_i, \mathrm{x}_j) = \exp(-\left\|\mathrm{x}_i - \mathrm{x}_j\right\|^2/(2\sigma)^2)$. Given a kernel function κ, KFCM can be generally defined as the constrained minimization of a reformulation of the objective:

$$f_m(U;\kappa) = \sum_{j=1}^{c}\left(\sum_{i=1}^{n}\sum_{k=1}^{n}(u_{ij}^m u_{kj}^m d_\kappa(\mathrm{x}_i, \mathrm{x}_k))/2\sum_{l=1}^{n}u_{lj}^m\right) \tag{9.7}$$

where U is the $(n \times c)$ partition matrix, and $n \in \mathrm{N}^+$ is the number of training data, $c \in [2, n)$ is the number of clusters, $m > 1$ is the fuzzification parameter, $d_\kappa(\mathrm{x}_i, \mathrm{x}_k) = \kappa(\mathrm{x}_i, \mathrm{x}_i) + \kappa(\mathrm{x}_k, \mathrm{x}_k) - 2\kappa(\mathrm{x}_i, \mathrm{x}_k)$ is the kernel-based distance between the ith and kth feature vectors. Note that in (9.7) we use a partition matrix that is $(n \times c)$ to stay consistent with the existing kernel clustering literature.

In the proposed AWKFCM, to define the relative importance of each cluster and each information, e.g., age, gender, and nationality, the adjusted weight is introduced into KFCM, and (9.7) is renewed as

$$f_m(U;\kappa) = \sum_{j=1}^{c}\left(\frac{\sum_{i=1}^{n}\sum_{k=1}^{n}((aw^c)_i u_{ij}^m (aw^c)_k u_{kj}^m d_\kappa(\mathrm{x}_i, \mathrm{x}_k))}{2\sum_{l=1}^{n}(aw^c)_l u_{lj}^m}\right) \tag{9.8}$$

where $aw^c \in \mathrm{R}^n$, $aw^c \geq 0$, is a set of adjusted weights that define the relative importance of each cluster c.

First, data clustering of the input information. Before data clustering, to indicate relative importance of each input information, e.g., age, gender, and nationality, x is renewed as

$$\mathrm{x}_i = (aw^{\mathrm{x}})_k \times Normalization(\mathrm{x}_i) \tag{9.9}$$

where $Normalization(\cdot)$ means normalization, and $aw^{\mathrm{x}} \in \mathrm{R}^n$, $(aw^{\mathrm{x}})_k \geq 0$ is a set of adjusted weights that defines the influence of each input information, which is used to emphasize the most important one, in normal case, $(aw^{\mathrm{x}})_k = 1$, and aw^{x} can be obtain by trial and error.

AWKFCM solves the optimization problem $\min\{f_m(U;\kappa)\}$ by computing iterated updates of

$$\mu_{ik} = \left(\sum_{k=1}^{c}\left(\frac{d_\kappa(\mathrm{x}_i, \mathrm{v}_j)}{d_\kappa(\mathrm{x}_i, \mathrm{v}_k)}\right)^{\frac{1}{(m-1)}}\right)^{-1} \quad \forall i, j. \tag{9.10}$$

Adjusted weights of each cluster is calculated as

$$(aw^c)'_j = \sum_{i=1}^{n}(u_{ij})(aw^c)_i. \tag{9.11}$$

The kernel distance between input datum x_i and cluster center v_j is

$$d_\kappa(x_i, v_j) = \left\|\phi(x_i) - \phi(v_j)\right\|^2 \tag{9.12}$$

where the cluster centers are linear combinations of feature vectors,

$$\phi(v_j) = \frac{\sum_{l=1}^{n}(aw^c)_l u_{lj}^m \phi(x_l)}{\sum_{l=1}^{n}(aw^c)_l u_{lj}^m}. \tag{9.13}$$

Equation (9.9) cannot be directly computed, and expanded into

$$\begin{aligned} d_\kappa(x_i, v_j) = & \frac{\sum_{l=1}^{n}\sum_{s=1}^{n}(aw^c)_l u_{lj}^m (aw^c)_s u_{sj}^m \phi(x_l)\cdot\phi(x_s)}{\sum_{l=1}^{n}(aw^c)_l^2 u_{lj}^{2m}} \\ & + \phi(x_i)\cdot\phi(x_i) - 2\frac{\sum_{l=1}^{n}(aw^c)_l u_{lj}^m \phi(x_l)\cdot\phi(x_i)}{\sum_{l=1}^{n}(aw^c)_l u_{lj}^m}. \end{aligned} \tag{9.14}$$

The iteration will stop when criterion is satisfied, i.e., $\left\|\phi(x_i) - \phi(v_j)\right\| \leqslant \phi(\varepsilon)$, where ε is the given sensitivity threshold.

Then the spread width is calculated as

$$\delta_j = \sqrt{\frac{\sum_{l=1}^{n}(aw^c)_l u_{lj}^m d_\kappa(x_i, v_j)}{\sum_{l=1}^{n}(aw^c)_l u_{lj}^m}}. \tag{9.15}$$

According to the obtained centers and spread width, the training data sample is split into training subsets as

$$D_j = \left\{\phi(v_j) - \eta\delta_j \leqslant \phi(x_i) \leqslant \phi(v_j) + \eta\delta_j\right\} \tag{9.16}$$

where η is a constant for controlling the overlap region of the training subsets and the size of training subsets increases as η increases, as well as computational time.

Second, understanding of the input information regression function of each cluster is constructed by (9.5) as

$$SVR_j = \sum_{i=1}^{n_j} \left(\alpha_{i,j}^* - \alpha_{i,j}\right) \kappa\left(\mathrm{x}_i, \mathrm{x}_k\right) + b_j, \mathrm{x} \in D_j \tag{9.17}$$

where n_j denotes the number of training data in the jth training subset, the parameters $\alpha_{i,j}^*$, $\alpha_{i,j}$ and b_j are obtained by the SVR for the jth training subset.

Finally, after information understanding, intention understanding for output is calculated based on the fuzzy weighted average algorithm [32] fusion model with triangle membership functions. The membership function of the fuzzy weighted average algorithm is built as

$$A_j(\mathrm{x}_i) = \max\left(\min\left(\frac{\phi(\mathrm{x}_i) - (\phi(\mathrm{v}_j) - \eta\delta_j)}{\phi(\mathrm{v}_j) - (\phi(\mathrm{v}_j) - \eta\delta_j)}, \frac{(\phi(\mathrm{v}_j) + \eta\delta_j) - \phi(\mathrm{x}_i)}{(\phi(\mathrm{v}_j) + \eta\delta_j) - \phi(\mathrm{v}_j)}\right), 0\right) \tag{9.18}$$

where $A_j(\mathrm{x}_i)$ is the fuzzy weight of the jth SVR.

Then, the output of the proposed TLFSVR is based on the fusion model which is calculated using fuzzy weighted average algorithm as

$$y(\mathrm{x}_i) = \frac{\sum_{j=1}^{c} A_j(\mathrm{x}_i) SVR_j(\mathrm{x}_i)}{\sum_{j=1}^{c} A_j(\mathrm{x}_i)}. \tag{9.19}$$

9.4 Characteristics Analysis of Emotional Intention Understanding

Intention are not directly observable, but can be inferred from expressive behavior, self-report, physiological indicators, and context. The ability to understand the intent of others is critical for the successful communication and collaboration between people. In our daily interactions, we rely heavily on the skill to "read" others' minds. If robots become effective in collaborating with humans, their cognitive skills must include mechanisms for inferring intention.

One way that humanoid robot facilitates intention understanding is based on emotion. Emotion helps inform and motivate social decision making, as well as the quality of our social relationships [33]. Our reactions to events indirectly convey information about the evaluation of current desires and intentions, e.g., an undesired stimulus might result in an expression of disgust and an unexpected one might result in an expression of surprise. Moreover, emotion expression seems to reflexively elicit adaptive social responses from others, and it can further trigger behavioral responses. For example, anger can elicit fear related responses or be served as a demand for someone to change their intention, sadness can elicit sympathy, and happy seems to invite social interaction.

Emotional intention understanding is defined as estimating customers' order intentions according to their emotion at a bar in our proposal, and emotion mainly consists of facial expression. Bar is a very popular place for both young people and salary man. At a bar, "what do you want to drink" is one of the customer's intention. If the bartender knows very well about what is the right drink of customer in different emotions, it would be very nice service to the customer and they may come back again.

Based on the data collected via brainstorming from twenty-eight volunteers (i.e., different genders people aged 20–65 years old, from 6 countries) in our previous study [37], age, gender, nationality, and emotion, are chose as the inputs of the intention understanding model by using hierarchical clustering algorithm and analysis of variance. Age is selected since not only it is the accumulation of changes (e.g., education, occupation, and income) in a person over time, but also some legal age are existed, such as drinking age, driving age, voting age, and so on; gender refers to female and male, it is selected because a broader set of hobby can be obtained from the gender point of view; nationality is selected because it indicates the legal relationship between a person and a state, and the person should obey the state law, such as people are illegal to drink alcohol in some Muslim country, in such a way that, nationality reflects religion. Moreover, age, gender, and nationality play important roles in intention understanding, for example, they affect people's intention to purchase products or services [34], accept new things [35], and handle risky situation [36]. With these in mind, three components of human property, i.e., age, gender, and nationality, are used for intention understanding beside emotion.

9.5 Three-Layer Fuzzy Support Vector Regression-Based Intention Understanding Model

Based on the TLWFSVR and characteristics analysis, an intention understanding model is proposed to comprehending peoples' inner thoughts in human-robot interaction. The understanding model reflects empirical findings accumulated in cognitive, at the same time, machine learning and fuzzy logic indicate the relationship of human intention and emotion. It is mainly consists of 8 steps:

Step 1: Initialization of AWKFCM clustering

Parameters, i.e., the number of clusters c, the sensitivity threshold ε, the overlap constant η, and adjusted weights of each information $(aw^{\mathrm{x}})_k$, are set.

Step 2: AWKFCM calculation

The membership u_{ik} is calculated by (9.8), the adjusted weights of each cluster $(aw^{c})_k$ is obtained by (9.11), the cluster centers v_j is calculated by (9.13), and the spread width δ_j is calculated by (9.10).

Table 9.1 Look-up table for output

Age	Nationality	
	Islam country or religion	Others
$<A_{lower}$	Non-alcoholic drinks	Non-alcoholic drinks
$A_{lower} \sim A_{upper}$	Non-alcoholic drinks	$I_{TLWFSVR}$
$>A_{upper}$	Non-alcoholic drinks	Non-alcoholic drinks

Step 3: If the stop criterion is not satisfied, then go to *Step* 2; otherwise, go to *Step* 4.

Step 4: AWKFCM clustering

The training subsets D_j are obtained by (9.11) based on the cluster centers and spread width.

Step 5: Initialization of SVR

Parameters, i.e., the Gaussian kernel width σ and the scalar regularization parameter γ are set.

Step 6: SVR learning

Regression function of each cluster SVR_j is constructed by SVR learning approach.

Step 7: Information understanding

Information understanding of the TLWFSVR is outputted base on the fuzzy weighted average algorithm by using (9.14).

Step 8: Intention understanding

Base on the identification information, i.e., age, gender, and nationality, output of intention understanding model is designed by the look-up table, as shown in Table 9.1, where A_{lower} is the legal drinking age according to different countries, e.g., 18 in China, 20 in Japan, 19 in South Korea, and so on. A_{upper} is the healthy drinking age, e.g. low-risk drinking style for healthy life begins from age 65 [37].

9.6 Experiments

To validate the proposed intention understanding model, simulation experiments and application experiments are both done. They choose the scenario of "drinking at a bar" as the background, this is due to the prevalence of the bar culture in the world, especially for young people and salary man. If the owner of a bar understands each customer preferences, e.g., a right drink for specific mood, it would be very nice and deeply touches the tired salary man, by which it will also increase the turnover. The goal of experiments is to understand customer's intention to order at a bar based on the customer's emotion.

Table 9.2 Questionnaire for relationship between emotion and intention to order at a bar

Questions	Answers
What do you want to order when your emotion is E? ($E \in$ {anger, disgust, sadness, surprise, fear, happiness, etc.}, 25 questions corresponding to 25 kinds of emotions in total, as shown in Fig. 9.3)	1-Wine, 2-beer, 3-sake, 4-sour, 5-shochu, 6-whisky, 7-non-alcoholic drink, 8-others (other drinks or food)

9.6.1 Experiment Setting

Simulation experiments are designed to verify the accuracy of the proposal, and conducted in the Matlab environment, which is ran in a PC with dual core processer (2.0 GHz and 2.6 GHz), memory (8 GB), and Windows 8 system. Two databases come from self-built and UCI Machine Learning Repository. Experiments are set based on the k-fold cross validation ($k = 10$), while one of the folds is used for testing, others are used for training. The preliminary application experiments are developed to confirm the validity and practicability of the proposed TLWFSVR model by using human-robot interaction system, called emotional social robot system, which is the new development of Mascot Robot System (MRS) [38].

9.6.2 Experiments on Three-Layer Fuzzy Support Vector Regression Based Intention Understanding Model

Two kinds of databases are prepared for the experiments. One is the self-built database, and the other is the benchmark database from UCI Machine Learning Repository [39], which is created as by David Aha and fellow graduate students at UC Irvine.

9.6.2.1 Data Preparation

In the self-built databases, data preparation is designed based on the scenario of "drinking at a bar". Data sample for simulation experiments is collected from sixty-two volunteers via questionnaires. Questionnaires are about the relationship between emotion and intention to order at a bar, at the same time, write down the identification information, such as age, gender, nationality, as shown in Table 9.2. The order intentions mainly include the top six popular drinks in the bar. To easier obtain data sample, questionnaires are transformed into information acquisition system in our previous study [40] as shown in Fig. 9.2, consisting of basic information acquisition (i.e., identification information and order intention selection) and emotion recognition.

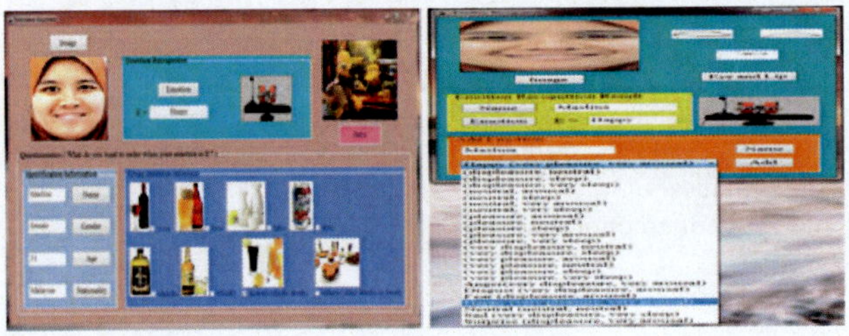

Fig. 9.2 Information acquisition system for self-built database

Table 9.3 Analysis of variance for self-built database

Source	Sum Sq.	d.f.	Mean Sq.	F	$P_{rob} > F$
Emotion	900.29	24	37.512	10.67	3.716×10^{-37}
Age	106.92	1	106.924	30.40	4.129×10^{-8}
Gender	1194.14	13	91.857	26.12	7.241×10^{-58}
Nationality	1197.26	4	299.315	85.1	3.197×10^{-65}
Error	5296.95	1506	3.517		
Total	8456.93	1549			

There are 62 volunteers, and they consist of 17 salary men, 25 postgraduate students, and 20 actual customers. 17 salary men aged 30 to 65 years old and 25 postgraduate students aged 20 to 35 old. They are living in Wuhan (China) and Tokyo (Japan) are of different genders, and come from eight countries, i.e., China, Japan, South Korea, Cuba, Mexico, and Muslim countries including Malaysia, Turkey, and Jordan. Moreover, to get more valid results, more data are collected by asking 20 actual customers at the bar named Suhe in Wuhan and the Izakaya named Watami in Tokyo, they are of age 20–60 years, including 12 male and 8 female, mainly from China and Japan. They ordered randomly, and if it is beyond the popular six drinks as shown in Table 9.3, the intention will be collected as 7-non-alcoholic drink or 8-others (other drinks or food). A total of 1550 groups of data are collected from sixty-two volunteers (17 salary men+25 postgraduate students+20 actual customers), and data of each group are recorded as {emotion, gender, age, national, intention}.

N-way ANOVA is used to examine the influence of emotion/age/gender/nationality on intention by using the collected data, where the independent variables are emotion, age, gender, and nationality, and the dependent variable is intention. The null hypothesis is all the emotion/age/gender/nationality groups have equal influence on the intention. According to analysis of variance in Table 9.3, all the P-values in 4 independent variables are lower than 0.05, thus the null hypothesis can be rejected, that means emotion, age, gender, and nationality have signification influence on the intention. Since the collected data can show the statistical significance of emotion, age, gender, nationality, they can be applied to the simulation.

The benchmark databases from UCI Machine Learning Repository called sentence classification data set, which contains sentences from the abstract and introduction of 30 articles in the area of biology, machine learning, and psychology.

9.6.2.2 Performance of Three-Layer Fuzzy Support Vector Regression Using Self-built Databases

To validate the performance of the proposed TLWFSVR, two sets of experiments are performed. The first experiments compare the performance of AWKFCM, KFCM, and FCM on data clustering of emotion-identification information. In the second sets of experiments, a comparative experiment among the TLWFSVR, TLFSVR, SVR, and BPNN is carried out, aiming to generate the intention by emotion and discuss the influence of age/gender/nationality on the intention understanding.

In the first set of experiments, Adjusted Rand Index (ARI) [41] and computational time (CT) are used to judge the performance of clustering, where ARI is a measure of agreement between two Boolean partitions of a set of objects. A Rand index of 1 indicates perfect agreement, while a Rand index of 0 indicates totally disagreement.

In the second set of experiments, four indexes are used to evaluate the performance, including accuracy, computational time, average absolute relative deviation (AARD), and correlation coefficient. AARD is defined as

$$\%AARD = \frac{100}{n} \sum_{i=1}^{n} \left| \frac{y_{cal} - y_{exp}}{y_{exp}} \right| \tag{9.20}$$

where n denotes the number of the testing data, y_{cal} represents the output of the proposal, y_{exp} is the actual value from information acquisition system.

The first experiments consist of three cases. In the fist case, the number of clusters c for the AWKFCM, KFCM, and FCM is chosen as 2 according to the gender (i.e., female and male). In the second case, the number of clusters c is chosen as 3 according to the age distribution (in general, people aged $\leq$30 belong to student, aged 30–50 belong to worker, and age $\geq$50 belong to retire). In the third case, the number of clusters c is chosen as 6 according to the nationality (volunteers are from 8 different countries in which 3 Muslim countries are in the same cluster).

Based on the experimental experience in our previous study [42], parameters of clustering methods are obtained by trial and error. For parameters setting of FCM, in general, the overlap parameter η can be selected as 2.5, 3.5, 4, and 5 corresponding to the dimension of input variables are 1, 2, 4, and 8, respectively. Since the input variables are 4 in the experiment, η can be 3 and the sensitivity threshold ε are set to 1. For KFCM, the Gaussian kernel width σ is suggested as $(0.1 \sim 0.5) \times$ (rang of the inputs), since the maximum area of inputs is the range of age (normally from 0 to 150), σ is chosen as 75. For AWKFCM, since the inputs are all normalized, the Gaussian kernel width σ is chosen as 0.5.

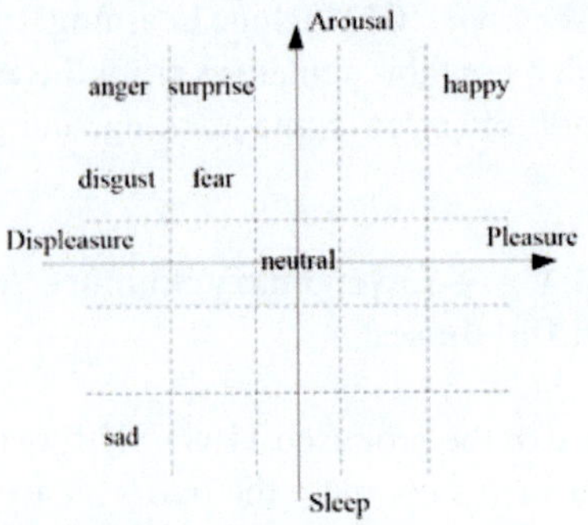

Fig. 9.3 Pleasure-arousal emotion plan

Figure 9.3 shows the clustering results of the proposed AWKFCM, KFCM, and FCM with $c = 2/3/6$, respectively, and the data analysis of result comparison among the three algorithms are showed in Table 9.4. According to the clustering results, while $c = 2/3/6$, the average ARI (in k-folds, $k = 10$) of the proposed AWKFCM is 1/0.9562/1, which is 85.63%/70.57%/25.32% and 157.93%/88.67%/74.70% higher than that of the KFCM and FCM, respectively. Average computational time (in k-folds, $k = 10$) is about 0.0048 s/0.0066 s/0.0011 s by using the proposed AWKFCM, indicating that the computing speed is about 3.55/1.59/1.85 and 1.81/1.38/2.17 times faster than that of the KFCM and FCM, respectively.

Moreover, it shows the change of the adjusted weights (normalized from 0 to 1) in the proposed AWKFCM, and the convergence of data clustering in the case of $k = 10$. According to the results, while $c = 2/3/6$, the iterations of the proposed AWKFCM is 3/17/12, which is 12/23/−2 and 14/6/21 iterations less than that of the KFCM and FCM, respectively. In the proposed AWKFCM, as the iterations increasing, adjusted weights of each cluster become smaller, and each cluster is classified more explicit. Each cluster is successfully distinguished one by one, in the case of cluster with more data than others, the adjusted weight of this cluster will be kept high enough for convergence. By introducing the adjusted weights into our proposal, there are two advantages. One is easier to convergence (less iterations) with the appropriate change of the adjusted weights. The other is much more clear classification can be obtained in each cluster. Without the change of the adjusted weights, each cluster easily overlaps. For example, while clustering using KFCM in the case of $c = 6$, even the iterations is 2 times less than the proposal, but the edge of each cluster is unclear. The data from Muslim Countries (i.e., Malaysia, Turkey, and Jordan) overlap with the data from South Korea, and the same case happens in Mexico and Cuba (Fig. 9.4).

In the second set of experiments, for parameters setting of each SVR, the Gaussian kernel width σ is suggested as $(0.1 \sim 0.5) \times$ (rang of the inputs), since the inputs are all normalized, σ is chosen as 0.5, and the scalar regularization parameter γ is chosen as 200. BPNN has three layers, since numbers of input notes and output notes are 4 (emotion, age, gender, and nationality) and 1 (intention), numbers of hidden

Table 9.4 Comparison result of information clustering with self-built database

k-fold	Index	Algorithm		
		AWKFCM	KFCM	FCM
		$c = 6/c = 3/c = 2$	$c = 6/c = 3/c = 2$	$c = 6/c = 3/c = 2$
k = 1	ARI	1.0000/0.9562/1.0000	0.8671/0.5606/0.5448	0.5391/0.5386/0.3152
	CT(s)	0.0042/0.0047/0.0016	0.0070/0.0102/0.0033	0.0116/0.0095/0.0017
k = 2	ARI	1.0000/0.9562/1.0000	0.7989/0.5606/0.5448	0.5391/0.4021/0.3192
	CT(s)	0.0064/0.0099/0.0009	0.0081/0.0087/0.0035	0.0093/0.0070/0.0018
k = 3	ARI	1.0000/0.9562/1.0000	0.7170/0.5606/0.5331	0.5391/0.4233/0.3029
	CT(s)	0.0048/0.0047/0.0008	0.0071/0.0189/0.0043	0.0085/0.0082/0.0021
k = 4	ARI	1.0000/0.9562/1.0000	0.6354/0.5606/0.4949	0.5391/0.4495/0.4581
	CT(s)	0.0039/0.0079/0.0028	0.0085/0.0098/0.0036	0.0136/0.0104/0.0018
k = 5	ARI	1.0000/0.9562/1.0000	0.6354/0.5606/0.6430	0.7074/0.5107/0.3624
	CT(s)	0.0041/0.0066/0.0009	0.0119/0.0127/0.0042	0.0072/0.0097/0.0018
k = 6	ARI	1.0000/0.9562/1.0000	0.8617/0.5606/0.6219	0.5391/0.7044/0.3059
	CT(s)	0.0047/0.0051/0.0008	0.0074/0.0119/0.0031	0.0131/0.0091/0.0019
k = 7	ARI	1.0000/0.9562/1.0000	0.7170/0.5606/0.5100	0.5391/0.4043/0.4654
	CT(s)	0.0076/0.0063/0.0008	0.0133/0.0065/0.0033	0.0108/0.0091/0.0020
k = 8	ARI	1.0000/0.9562/1.0000	0.7170/0.5606/0.4438	0.5391/0.4979/0.3037
	CT(s)	0.0042/0.0081/0.0009	0.0058/0.0114/0.0032	0.0098/0.0095/0.0020
k = 9	ARI	1.0000/0.9562/1.0000	0.7973/0.5606/0.5138	0.7034/0.5916/0.6388
	CT(s)	0.0032/0.0090/0.0007	0.0121/0.0044/0.0069	0.0108/0.0093/0.0027
k = 10	ARI	1.0000/0.9562/1.0000	0.6354/0.5606/0.5370	0.5391/0.5071/0.4055
	CT(s)	0.0047/0.0041/0.0008	0.0087/0.0103/0.0035	0.0090/0.0092/0.0018
Average	ARI	1.0000/0.9562/1.0000	0.7979/0.5606/0.5387	0.5724/0.5068/0.3877
	CT(s)	0.0048/0.0066/0.0011	0.0089/0.0105/0.0039	0.0104/0.0091/0.0020

notes can be 12 according to the empirical formula $\sqrt{input\ notes{+}output\ notes} + a$ ($1 \leqslant a \leqslant 10$), where learning rate is 0.25 and inertia coefficient is 0.05.

Figure 9.5 shows the intention understanding output of the proposed TLWFSVR, TLFSVR, SVR, and BPNN with $c = 2/3/6$, respectively, and the data analysis of result comparison among the three methods is showed in Table 9.5. According to simulation results, while $c = 2/3/6$, the proposed TLWFSVR obtains an average accuracy (in k-folds, $k = 10$) of 64.13%/68.31%/82.32%, which is 11.04%/12.38% /13.51%, 19.04%/23.22%/37.23%, and 33.92%/38.10%/52.11% higher than that of TLFSVR, SVR, and BPNN, respectively. Average Computational time (in k-folds, $k = 10$) is about 0.25 s/0.30 s/0.31 s by using the proposal, indicating that the computing speed is about 1.08/1.11/1.03, 7.76/7.19/6.26 and 14.36/13.29/11.58 times faster than the TLFSVR, SVR, and BPNN, respectively.

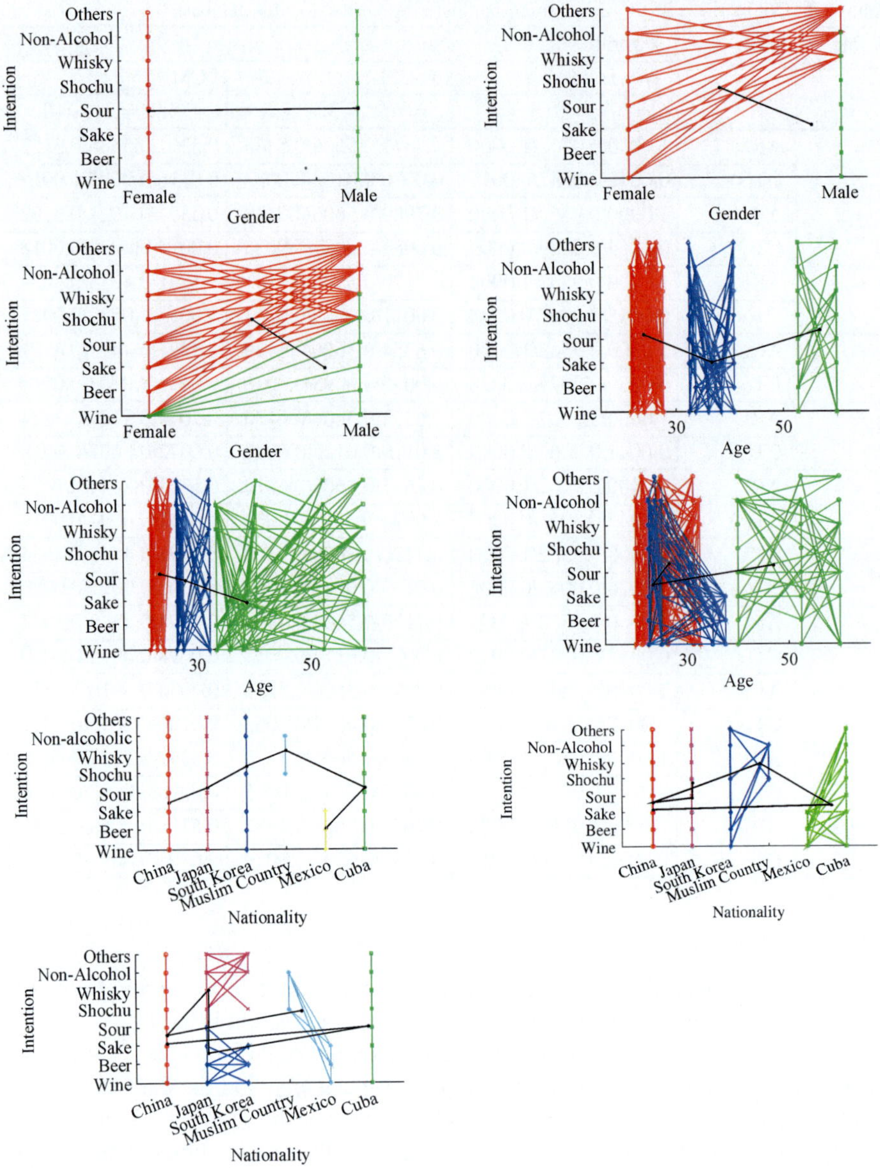

Fig. 9.4 Data clustering based on gender/age/nationality (in the case of $k = 10$)

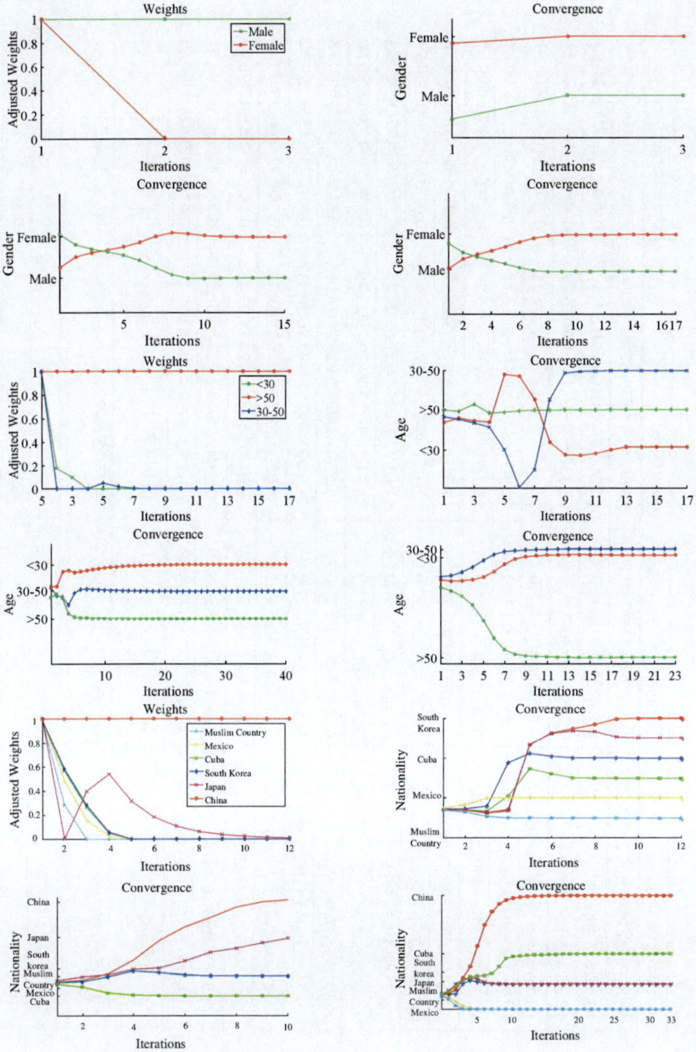

Fig. 9.5 Data clustering based on gender/age/nationality (in the case of $k = 10$)

9.6.2.3 Performance of Three-Layer Fuzzy Support Vector Regression Using UCI Database

Another database from UCI Machine Learning Repository [43] is used to validate the benchmark of the proposed TLWFSVR. In the experiments, 1280 sentences are selected from the sentence classification data set in the UCI repository, where 4 types of sentences include “1-Own sentences”, “2-Contrary sentences”, “3-Base sentences”, and “4-Aim sentences”. Based on the word frequency in each type of sentences, key words of each type are selected and shown in Table 9.6.

Table 9.5 Comparison result of intention understanding with self-built database

Algorithm		Index			
		Accuracy (%)	Computational time (s)	AARD (%)	Correlation
TLWFSVR	$c = 6$	83.87/85.16/79.35/81.29/83.23	0.29/0.29/0.32/0.31/0.31	9.76/7.42/11.82/9.12/8.54	0.94/0.92/0.92/0.94/0.94
		84.52/81.94/82.58/80.00/81.29	0.30/0.31/0.33/0.31/0.32	8.73/11.54/10.84/9.24/12.42	0.96/0.92/0.92/0.94/0.93
	$c = 3$	70.32/68.93/67.10/69.68/63.87	0.25/0.26/0.30/0.25/0.26	13.43/15.06/16.17/14.94/16.29	0.89/0.86/0.87/0.88/0.84
		68.39/67.74/66.45/71.61/69.03	0.26/0.29/0.26/0.25/0.26	14.38/16.39/15.71/12.24/14.71	0.86/0.83/0.85/0.88/0.87
	$c = 2$	65.81/64.52/60.65/65.16/68.39	0.25/0.28/0.23/0.25/0.24	13.90/14.61/18.65/15.25/13.28	0.86/0.88/0.86/0.87/0.89
		63.87/63.23/61.94/61.29/66.45	0.24/0.23/0.30/0.25/0.24	13.28/14.63/16.16/16.39/15.37	0.88/0.88/0.86/0.86/0.87
TLFSVR	$c = 6$	71.55/72.19/66.45/71.61/64.52	0.30/0.32/0.28/0.32/0.30	11.31/10.54/16.82/17.78/22.17	0.92/0.90/0.85/0.86/0.76
		73.48/76.35/63.23/68.39/70.26	0.37/0.32/0.37/0.32/0.29	9.81/11.49/21.77/12.40/10.38	0.92/0.91/0.76/0.88/0.90
	$c = 3$	56.13/54.84/52.90/57.42/53.55	0.28/0.28/0.31/0.32/0.29	21.03/22.16/25.14/21.69/26.45	0.82/0.76/0.77/0.78/0.75
		60.00/55.48/56.77/58.06/54.19	0.29/0.32/0.34/0.31/0.29	24.11/21.78/23.70/20.75/23.76	0.75/0.79/0.78/0.80/0.80
	$c = 2$	56.77/53.55/49.03/46.45/55.48	0.26/0.27/0.32/0.26/0.27	19.92/21.18/23.53/25.94/18.06	0.74/0.74/0.75/0.69/0.78
		59.35/58.06/45.81/52.26/54.19	0.28/0.27/0.27/0.27/0.27	21.90/21.46/25.46/22.26/20.51	0.75/0.75/0.65/0.73/0.72
SVR		45.81/43.87/45.16/42.58/40.65	2.01/1.96/1.90/1.93/1.94	30.32/38.69/33.72/38.29/38.12	0.78/0.72/0.66/0.71/0.63
		48.93/44.52/42.58/50.32/46.45	1.88/1.99/1.90/1.88/1.96	28.04/40.03/41.14/26.87/37.45	0.78/0.67/0.66/0.77/0.69
BPNN		28.39/29.86/31.61/29.03/29.68	3.69/3.55/3.68/3.54/3.49	41.53/42.75/42.60/37.35/36.84	0.51/0.44/0.68/0.64/0.61
		31.61/29.68/29.03/30.97/32.26	3.70/3.50/3.71/3.54/3.53	37.78/42.33/37.46/34.83/35.19	0.67/0.59/0.62/0.57/0.64
Average	TLWFSVR ($c = 6/3/2$)	82.32/68.31/64.13	0.31/0.27/0.25	9.94/14.93/15.15	0.93/0.86/0.87
	TLFSVR ($c = 6/3/2$)	69.81/55.93/53.09	0.32/0.30/0.27	14.45/23.06/22.02	0.87/0.78/0.73
	SVR	45.09	1.94	35.27	0.71
	BPNN	30.21	3.59	38.87	0.59

Table 9.6 Key words of 4 types of sentences

Type	Key works
1-Own sentences	We, this, model, are, have, section, result, use, algorithm, study, show, propose
2-Contrary sentences	Difficult, however, only, many, may, unlike, relative, general, even, must, limit, while
3-Base sentences	Use, extend, relay, spirit, previous, foundation, on, reuse, similar, legacy, refine, early
4-Aim sentences	We, this, chapter, study, who, present, new, model, investigate, how, use, approach

Experiments and parameters setting are done as the same in the self-built databases besides two parts. Fist, the number of clusters c for the AWKFCM, KFCM, and FCM is chosen as 2 according to the feature key word "We" (i.e., with and without); second, in BPNN, since numbers of input notes and output notes are 48 (features in Table 9.6) and 1 (sentence meaning), numbers of hidden notes can be 17 according to the empirical formula $\sqrt{input\ notes{+}output\ notes} + a(1 \leqslant a \leqslant 10)$. The goal is to understand the meanings of each sentence into 4 types based on the features shown in Table 9.6.

Figure 9.5 shows the clustering results of the proposed AWKFCM, KFCM, and FCM with $c = 2$, respectively, and the data analysis of result comparison among the three algorithms are showed in Table 9.7. According to the clustering results, while $c = 2$, the average ARI (in k-folds, $k = 10$) of the proposed AWKFCM is 1, which is 3.3% and 76.18% higher than that of the KFCM and FCM, respectively. Computational time is about 0.0009 s by using the proposed AWKFCM, indicating that the computing speed is about 1.67 and 1.22 times faster than that of the KFCM and FCM, respectively. Moreover, it shows the change of the adjusted weights (normalized from 0 to 1) in the proposed AWKFCM, and the convergence of data clustering in the case of $k = 10$. According to the results, the distance between each cluster is larger than that of the KFCM and FCM, in which clustering data are close in KFCM, even overlapped in FCM. By introducing the adjusted weights into our proposal, each cluster is classified more explicit.

Figure 9.6 shows the sentence meaning understanding of the proposed TLWFSVR, TLFSVR, SVR, and BPNN with $c = 2$, respectively, and the data analysis of result comparison among the three methods is showed in Table 9.8. According to simulation results, while $c = 2/3/6$, the proposed TLWFSVR obtains an average accuracy (in k-folds, $k = 10$) of 87.89%, which is 24.27%, 31.66%, and 39.77% higher than that of TLFSVR, SVR, and BPNN, respectively. Average Computational time (in k-folds, $k = 10$) is about 0.25 s by using the proposal, indicating that the computing speed is about 1.48, 17.40, and 21.32 times faster than the TLFSVR, SVR, and BPNN, respectively.

According to the experimental results of two databases, for data clustering, the proposed AWKFCM obtains highest ARI, indicating that it is the most efficient algo-

Table 9.7 Comparison result of information clustering ($c = 2$) with UCI database

k-fold	Index	Algorithm		
		AWKFCM	KFCM	FCM
k = 1	ARI	1.0000	0.9676	0.7352
	CT(s)	0.0011	0.0016	0.0012
k = 2	ARI	1.0000	0.9676	0.5779
	CT(s)	0.0010	0.0015	0.0011
k = 3	ARI	1.0000	0.9676	0.6526
	CT(s)	0.0009	0.0016	0.0010
k = 4	ARI	1.0000	0.9676	0.4927
	CT(s)	0.0011	0.0014	0.0012
k = 5	ARI	1.0000	0.9676	0.7776
	CT(s)	0.0010	0.0012	0.0010
k = 6	ARI	1.0000	0.9676	0.5726
	CT(s)	0.0010	0.0017	0.0011
k = 7	ARI	1.0000	0.9676	0.5923
	CT(s)	0.0012	0.0015	0.0012
k = 8	ARI	1.0000	0.9676	0.4298
	CT(s)	0.0009	0.0014	0.0011
k = 9	ARI	1.0000	0.9676	0.6970
	CT(s)	0.0009	0.0012	0.0011
k = 10	ARI	1.0000	0.9676	0.4485
	CT(s)	0.0010	0.0020	0.0012
Average	ARI	1.0000	0.9676	0.5676
	CT(s)	0.0009	0.0015	0.0011

rithm among these there. The AWKFCM can more clearly divide male and female, 3 age area ($\leq$30, 30–50, and $\geq$50), people from 6 areas (China, Japan, South Korea, Cuba, Mexico, and Muslim countries including Malaysia, Turkey, and Jordan) than other two algorithms, this is because the adjusted weight represents the significance of different identification information, which will effect the clustering preference, leading to more precise categorization. The computational time is affected by the computational complexity of the algorithm, which is related to problem variables: n is the number of objects in the d-dimensional data, $X \in R^d$; c is the number of clusters; t is the number of iterations required for termination. For AWKFCM and KFCM, the complexity can be represented as $O(tcn^2)$, and $O(tc^2dn)$ for FCM. It seems that the complexity of AWKFCM and KFCM is the same, even large than FCM, however, the most important impact factor is t, where approximated adjusted weight will make the algorithm convergence as soon as possible, and lead to less iterations so that the computational time of AWKFCM is less than that of the KFCM and FCM.

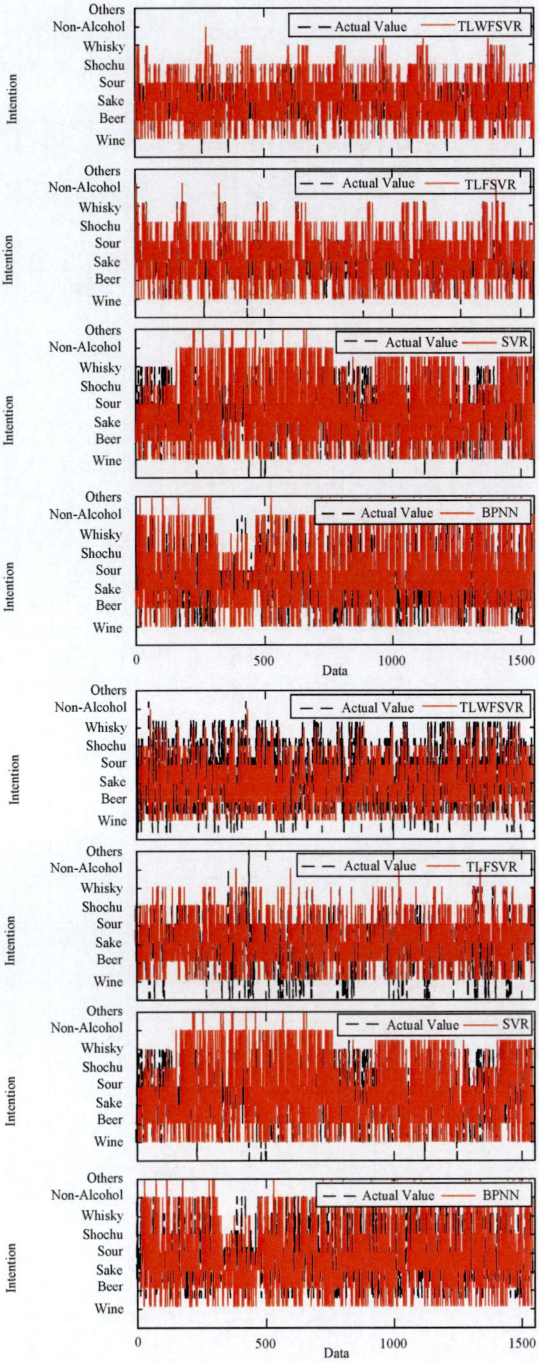

Fig. 9.6 Intention understanding with the use of TLWFSVR, TLFSVR, SVR, and BPNN

Table 9.8 Comparison result of intention understanding with UCI database

Algorithm		Index			
		Accuracy (%)	Computational time (s)	AARD (%)	Correlation
TLWFSVR	$c = 2$	93.70/92.19/82.81/89.06/82.03	0.24/0.25/0.25/0.25/0.26	1.84/2.21/7.42/3.06/5.60	0.98/0.97/0.82/0.97/0.92
		89.06/84.38/89.06/85.94/90.63	0.25/0.24/0.60/0.65/0.62	2.80/4.43/2.99/3.84/2.34	0.97/0.96/0.97/0.97/0.97
TLFSVR	$c = 2$	62.99/64.84/60.16/64.84/62.50	0.27/0.28/0.26/0.28/0.25	27.29/27.86/30.53/27.54/27.34	0.79/0.81/0.74/0.72/0.78
		64.94/60.94/61.54/61.72/67.19	0.27/0.26/0.27/0.27/0.26	27.77/28.58/27.79/26.17/21.61	0.78/0.71/0.83/0.76/0.82
SVR		54.33/62.50/54.69/60.30/52.34	4.43/4.68/4.48/4.25/4.15	36.55/29.95/35.42/33.86/36.72	0.78/0.82/0.74/0.73/0.78
		51.56/56.25/54.69/57.81/57.81	4.42/4.27/4.25/4.23/4.40	39.19//34.83/38.67/32.16/32.16	0.69/0.70/0.71/0.75/0.75
BPNN		63.78/57.81/47.66/50.78/10.06	5.31/5.29/5.37/5.20/5.34	28.81/36.39/45.83/42.71/99.02	0.20/0.04/0.05/0.02/0.13
		60.44/61.72/51.56/56.25/21.09	5.18/5.16/5.28/5.54/5.65	32.10/27.99/42.06/41.08/77.41	0.03/0.12/0.21/0.07/0.25
Average	TLWFSVR	87.89	0.25	3.65	0.95
	TLFSVR	63.17	0.37	27.25	0.77
	SVR	56.23	4.35	34.95	0.75
	BPNN	48.12	5.33	47.34	0.11

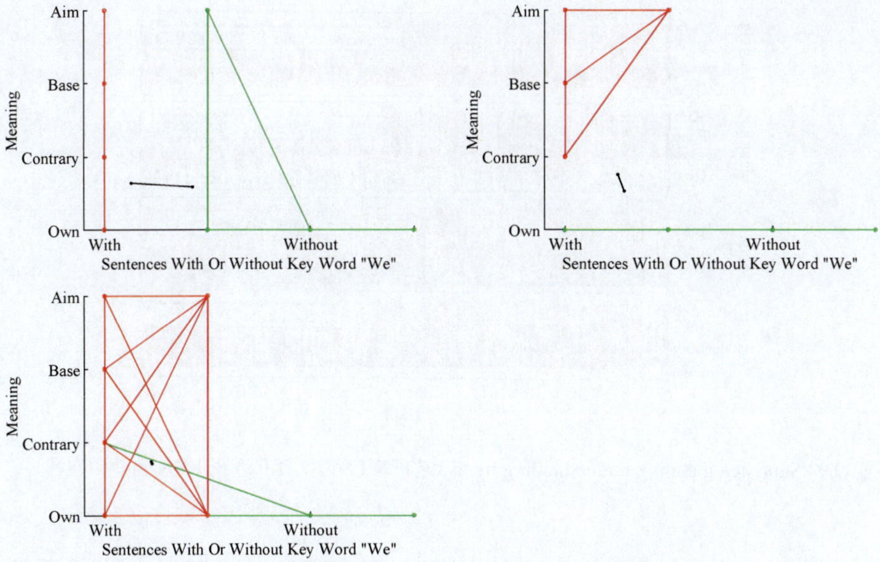

Fig. 9.7 Data clustering based on key word "We" ($c = 2$ in the case of $k = 10$)

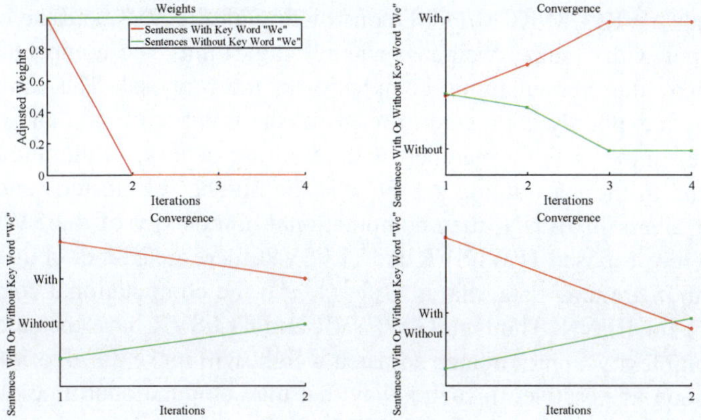

Fig. 9.8 Adjusted weights and convergence of data clustering based on key word "We" ($c = 2$ in the case of $k = 10$)

Moreover, it can be found that the accuracy of the proposed understanding model is higher than that of the TLFSVR, SVR, and BPNN, leading to the smaller AARD and higher correlation coefficient between the output and the actual value. Smaller AARD and higher correlation indicate that the TLFSVR based intention understanding model better matches the actual situation. The increase of cluster number c improves the understanding accuracy to some extent. The overlap parameter ε is an important parameter in the proposal, which becomes larger as the dimension of the

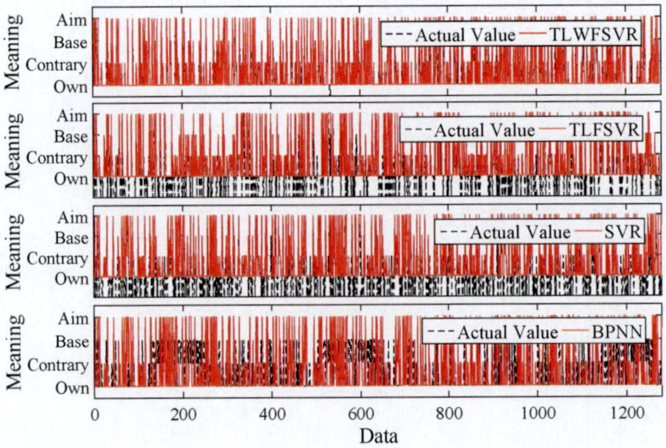

Fig. 9.9 Sentence meaning understanding by using TLWFSVR, TLFSVR, SVR, and BPNN

training data becomes higher. Since the overlap parameter ε can be set large to cover all the training data, the accuracy can be guaranteed.

The computational time mainly includes some processes such as data clustering by using AWKFCM/KCM/FCM, constructing of all SVRs, and the calculation of the output with a fuzzy weighted average algorithm. The computational time is affected by the computational complexity of the proposal. The computational complexity is explicitly expressed in terms of the number of data and computing scale, for example, n is the number of the training data, n_i is the number of the training data in the ith training subset. c is the number of clusters, and m is the number of layers in BPNN, then computational complexity of the SVR is set as $O(n^2)$; for the proposed TLWFSVR and TLFSVR, there are c times of the SVR conducted with d_i training data, that is $O(\sum_{i=1}^{C} n_i^2)$; the computational complexity is $O(m \times n^2)$ for BPNN. Thought TLWFSVR and TLFSVR have the same computational complexity, approximated adjusted weight will make the algorithm convergence as soon as possible, in such a way that the computational time of proposed TLWFSVR is less than that of the TLFSVR, SVR, and BPNN.

9.7 Summary

A three-layer weighted fuzzy support vector regression (TLWFSVR) based intention understanding model is proposed to fully comprehend people's inner thoughts. The key advantage of TLWFSVR is its learning ability to show the weights of each feature in each cluster, which indicates the importance of different HRI information, and endows robot with emotional intention understanding capability.

According to experimental results by using the self-built databases and benchmark databases from UC Irvine Machine Learning Repository, it can be found that the accuracy of the proposed understanding model is higher than that of the TLFSVR, SVR, and BPNN, leading to the smaller Adjusted Rand Index (AARD) and higher correlation coefficient between the output and the actual value. Smaller AARD and higher correlation indicate that the TLFSVR based intention understanding model better matches the actual situation. The clusters number c and the overlap parameter η are two important parameters in the proposed understanding model, which guarantees the accuracy. The computational complexity is explicitly expressed in terms of the number of data and computing scale, for example, n is the number of the training data, n_i is the number of the training data in the ith training subset. c is the number of clusters, and m is the number of layers in BPNN, then computational complexity of the SVR is set as $O(n^2)$; for the proposed TLWFSVR and TLFSVR, there are c times of the SVR conducted with d_i training data, that is $O(\sum_{i=1}^{c} n_i^2)$; the computational complexity is $O(m \times n^2)$ for BPNN. Thought TLWFSVR and TLFSVR have the same computational complexity, approximated adjusted weight will make the algorithm convergence as soon as possible, in such a way that the computational time of proposed TLWFSVR is less than that of the TLFSVR, SVR, and BPNN.

References

1. E. Noohi, M. Zefran, J.L. Patton, A model for human-human collaborative object manipulation and its application to human-robot interaction. IEEE Trans. Robot. **32**(4), 880–896 (2016)
2. K.K. Roudposhti, U. Nunes, J. Dias, Probabilistic social behavior analysis by exploring body motion-based patterns. IEEE Trans. Pattern Anal. Mach. Intell. **38**(8), 1679–1691 (2016)
3. K. Zheng, D.F. Glas, T. Kanda, H. Ishiguro, N. Hagita, Designing and implementing a human-robot team for social interactions. IEEE Trans. Syst. Man Cybern. Syst. **43**(4), 843–858 (2013)
4. K. Pitsch, T. Dankert, R. Gehle, S. Wrede, Referential practices. Effects of a museum guide robot suggesting a deictic 'repair' action to visitors attempting to orient to an exhibit, in *Proceedings of IEEE International Symposium on Robot and Human Interactive Communication*, NY, USA, pp. 225–231 (2016)
5. T. Kanda, M. Shiomi, Z. Miyashita, H. Ishiguro, N. Hagita, A communication robot in a shopping mall. IEEE Trans. Robot. **26**(5), 897–913 (2010)
6. J. Fasola, M. Mataric, Using socially assistive human-robot interaction to motivate physical exercise for older adults. Proc. IEEE **100**(8), 2512–2526 (2012)
7. J.E. Young, J. Sung, A. Voida, E. Sharlin, T. Igarashi, H.I. Christensen, R.E. Grinter, Evaluating human-robot interaction. Int. J. Soc. Robot. **3**(1), 53–67 (2011)
8. C. Zhang, H. Zhang, L.E. Parker, Feature space decomposition for effective robot adaptation, in *Proceedings of IEEE/RSJ International Conference on Intelligent Robots and Systems (IROS)*, Hamburg, Germany, pp. 441–448 (2015)
9. V.C. Meola, D. Caligiore, V. Sperati, L. Zollo, A.L. Ciancio, F. Taffoni, E. Guglielmelli, G. Baldassarre, Interplay of rhythmic and discrete manipulation movements during development: a policy-search reinforcement-learning robot model. Proc. IEEE Trans. Cogn. Dev. Syst. **8**(3), 152–170 (2016)
10. D. Hu, Y. Gong, B. Hannaford, E.J. Seibel, Semi-autonomous simulated brain tumor ablation with RAVENII surgical robot using behavior tree, in *Proceedings of IEEE International Conference on Robotics and Automation (ICRA)*, Seattle, USA, pp. 3868–3875 (2015)
11. A.M.C. Smith, C. Yang, H. Ma, P. Culverhouse, A. Cangelosi, E. Burdet, Novel hybrid adaptive controller for manipulation in complex perturbation environments. PLoS One **10**(6), e0129281 (2015)

12. S. Qiu, Z. Li, W. He, L. Zhang, C. Yang, C.Y. Su, Brain – machine interface and visual compressive sensing-based teleoperation control of an exoskeleton robot. IEEE Trans. Fuzzy Syst. **25**(1), 58–69 (2017)
13. L.F. Chen, Z.T. Liu, M. Wu, F.Y. Dong, Y. Yamazaki, K. Hirota, Multi-robot behavior adaptation to local and global communication atmosphere in humans-robots interaction. J. Multimodal User Interfaces **8**(3), 289–303 (2014)
14. Z. Li, C. Yang, C.Y. Su, S. Deng, F. Sun, W. Zhang, Decentralized fuzzy control of multiple cooperating robotic manipulators with impedance interaction. IEEE Trans. Fuzzy Syst. **23**(4), 1044–1056 (2015)
15. A. Geiger, M. Lauer, R. Urtasun, A generative model for 3D urban scene understanding from movable platforms, in *Proceedings of IEEE Conference on Computer Vision and Pattern Recognition*, Colorado, USA, pp. 1945–1952 (2011)
16. M.J. Roberson, J. Bohg, G.L. Skantze, J. Gustafson, R. Carlson, B. Rasolzadeh, D. Kragic, Enhanced visual scene understanding through human-robot dialog, in *Proceedings of IEEE/RSJ International Conference on Intelligent Robots and Systems*, San Francisco, USA, pp. 3342–3348 (2011)
17. M.V. Bergh, D. Carton, R.D. Nijs, N. Mitsou, C. Landsiedel, K. Kuehnlenz, D. Wollherr, L.V. Gool, M. Buss, Real-time 3D hand gesture interaction with a robot for understanding directions from humans, in *Proceedings of IEEE International Symposium on Robot and Human Interactive Communication*, Atlanta, USA, pp. 357–362 (2011)
18. C. Mohiyeddini, R. Pauli, S. Bauer, The role of emotion in bridging the intention behaviour gap: the case of sports participation. Psychol. Sport. Exerc. **10**(2), 226–234 (2009)
19. C. Cortes, V.N. Vapnik, Support-vector networks. Mach. Learn. **20**(3), 273–297 (1995)
20. H. Yu, J. Kim, Y. Kim, S. Hwang, Y.H. Lee, An efficient method for learning nonlinear ranking SVM functions. Inf. Sci. **209**, 37–48 (2012)
21. V.N. Vapnik, *The Nature of Statistical Learning Theory* (Springer, New York, 2000)
22. D. Song, D. Tao, Biologically inspired feature manifold for scene classification. IEEE Trans. Image Process. **19**(1), 174–184 (2010)
23. B. Andreas, A.W. Jamie, G. Hans, T. Gerhard, Eye movement analysis for activity recognition using electrooculography. IEEE Trans. Pattern Anal. Mach. Intell. **33**(4), 741–753 (2011)
24. W. Zhao, J. Zhang, K. Li, An efficient LS-SVM-based method for fuzzy system construction. IEEE Trans. Fuzzy Syst. **23**(3), 627–643 (2015)
25. L. Bottou, V.N. Vapnik, Local learning algorithms. Neutral Comput. **4**(6), 888–900 (1992)
26. C.F. Juang, C.D. Hsieh, A fuzzy system constructed by rule generation and iterative linear SVR for antecedent and consequent parameter optimization. IEEE Trans. Fuzzy Syst. **20**(2), 372–384 (2012)
27. X.X. Zhang, Y. Jiang, H.X. Li, S.Y. Li, SVR learning-based spatiotemporal fuzzy logic controller for nonlinear spatially distributed dynamic systems. IEEE Trans. Neural Netw. Learn. Syst. **24**(3), 1635–1647 (2013)
28. L.F. Chen, Z.T. Liu, M. Wu, M. Ding, F.Y. Dong, K. Hirota, Emotion-age-gender-nationality based intention understanding in human-robot interaction using two-layer fuzzy support vector regression. Int. J. Soc. Robot. **7**(5), 709–729 (2015)
29. M. Moavenian, H. Khorrami, A qualitative comparison of artificial neural networks and support vector machines in ECG arrhythmias classification. Expert Syst. Appl. **37**(4), 3088–3093 (2010)
30. K.P. Lin, A novel evolutionary kernel intuitionistic fuzzy c-means clustering algorithm. IEEE Trans. Fuzzy Syst. **22**(5), 1074–1087 (2014)
31. D.D. Nguyen, L.T. Ngo, Multiple kernel interval type-2 fuzzy c-means clustering, in *Proceedings of IEEE International Conference on Fuzzy Systems*, Hyderabad, India, pp. 1–8 (2013)
32. W.M. Dong, F.S. Wong, Fuzzy weighted averages and implementation of the extension principle. Fuzzy Sets Syst. **21**(2), 183–199 (1987)
33. D. Keltner, J. Haidt, Social functions of emotions at four levels of analysis. Cogn. Emot. **13**(5), 505–521 (1999)
34. A. Shafi, W. Vishanth, Understanding citizens' behavioral intention in the adoption of e-government services in the state of Qatar, in *Proceedings of European Conference on Information Systems*, Verona, Italy (2009)

35. R.O. Orji, Impact of gender and nationality on acceptance of a digital library: an empirical validation of nationality based UTAUT using SEM. J. Emerg. Trends Comput. Inf. Sci. **1**(2), 68–79 (2010)
36. C. Holland, R. Hill, The effect of age, gender and driver status on pedestrians' intentions to cross the road in risky situations intentions to cross the road in risky situations. Accid. Anal. Prev. **39**(2), 224–237 (2007)
37. G. Farchi, F. Fidanza, S. Giampaoli, S. Mariotti, A. Menotti, Alcohol and survival in the Italian rural cohorts of the seven countries study. Int. J. Epidemiol. **29**, 667–671 (2000)
38. K. Hirota, F.Y. Dong, Development of mascot robot system in NEDO project, in *Proceedings of International IEEE Conference on Intelligent Systems*, Varna, Bulgaria, pp. 38–44 (2008)
39. UCI Machine Learning Repository Data Sets, Sentence classification data set. UC Irvine, Irvine, CA, USA. Available: http://archive.ics.uci.edu/ml/datasets.html
40. L. Hubert, P. Arabie, Comparing partitions. J. Classif. **2**, 193–218 (1985)
41. L.F. Chen, M. Wu, M. Zhou, J. She, K. Hirota, Dynamic emotion understanding using FCM based SVR in human-robot interaction, in *Proceedings of the 35th Chinese Control Conference*, Chengdu, China, pp. 7064–7069 (2016)
42. J. Lafaye, C. Collette, P.B. Wieber, Model predictive control for tilt recovery of an omnidirectional wheeled humanoid robot, in *Proceedings of IEEE International Conference on Robotics and Automation*, Seattle, USA, pp. 5134–5139 (2015)
43. A.D. Ames, Human-inspired control of bipedal walking robots. IEEE Trans. Autom. Control **59**(5), 1115–1130 (2014)

Chapter 10
Dynamic Emotion Understanding Based on Two-Layer Fuzzy Fuzzy Support Vector Regression-Takagi-Sugeno Model

Two-layer Fuzzy SVR-TS Model is proposed for emotion understanding in human-robot interaction, where the real-time dynamic emotion is recognized according to facial expression, and emotional intention understanding is obtained mainly based on human emotions and identification information. It aims to make robots capable of recognizing and understanding human emotions, in such a way that make human-robot interaction run smoothly. TLFSVR-TS considers prior knowledge inferred from human personal preference to reduce the uncertainty of various people, and multiple SVRs corresponding to different genders/provinces/ages of human to guarantee the local learning ability.

10.1 Introduction

With the rapidly development of face recognition technology [1–3], as well as the extensive attention of many scholars, emotion recognition [4–6] is widely used as a way of affective computing [7, 8], and becomes an important part of human-robot interaction. Furthermore, fusion of facial expression recognition greatly promotes the understanding of human emotional state to achieve the smooth human-robot interaction [9–11].

Most of the researches focus on static image recognition [12–14], such as methods based on geometric features, wavelet transform method based on Gabor, local binary model algorithm, and method based on hybrid feature extraction. Al-Arashi et al. [12] proposed genetic algorithm [13] for facial expression recognition. Principal component analysis (PCA) combined with logistic regression analysis was presented to deal with image information. Even though static image recognition obtains high efficiency, speed, and other characteristics, some disadvantages of these methods are

L. Chen et al., *Emotion Recognition and Understanding for Emotional Human-Robot Interaction Systems*, Studies in Computational Intelligence 926,
https://doi.org/10.1007/978-3-030-61577-2_10

that they lack sufficient dynamic information, and are easy affected by environment and individual differences, such as the different complexion, facial features, and light intensity. However, the dynamic emotion recognition of real-time is important in human-robot interaction. There are some research on dynamic face [15–17], commonly including optical flow method [18], feature point tracking method [19, 20], motion model algorithm [21, 22], and others. Most of the existing algorithms focus on the sequence of dynamic image, in which real-time dynamic emotion recognition can not be directly realized and human emotion understanding is not discussion, [23] can be taken into account, where the facial expression recognition in image sequences is realized by using support vector machines (SVM) [24, 25] based on geometric deformation features. So that under this consideration, we use candide3 based feature point matching method, which can not only obtain the feature point, but also the dynamic information.

Studies are then moving on to obtain the solution for emotion understanding. At present, support vector regression (SVR) has been successfully used for prediction in different fields. For instance, Hu et al. [26] proposed heteroscedastic SVR, in which has been used for short-term forecasting of wind power or speed. Cheng et al. [27] identified a fuzzy model for classification, in which margin-selective gradient descent learning and online incremental SVM are used. In [28], SVR with an fuzzy-neural network was proposed and employed for noisy regression problems, in which the noise resistance of network is improved with the help of interval type-2 fuzzy sets, and IT2FNN-SVR is used for parameter learning and structure learning. Although SVR is powerful for prediction or classification, it cannot express causal relationship between input and output spaces. Fuzzy 'If-Then' rules-based fuzzy model' is introduced to show the causality. Takagi-Sugeno (TS) fuzzy model is linear system model to express local dynamics of fuzzy rules, and it receives many successful applications in nonlinear control systems [29–32]. For example, Cheung et al. [33] introduced parameters searching strategy by using TS fuzzy system with PSO to enhance the searching performance. Chang et al. [34] developed a TS fuzzy model combined with SVR for stock trading forecasting.

As a matter of fact, human-robot interaction (HRI) system can also be considered as a nonlinear system, and fuzzy model can be used. Moreover, intention understanding can be regarded as a prediction or classification problem. In our previous study [35], Two-Layer Fuzzy SVR (TLFSVR) was proposed for intention understanding in human-robot interaction, in which fuzzy c-means is used for data classification and SVR is used for learning. In [36], a three-domain fuzzy support vector regression is provided for image understanding in human-robot interaction, which focused on mapping the input and output domain into the fuzzy domain. Similar to our previous study TLFSVR in [35], kernel fuzzy c-means clustering is fused with SVR (KFCM-FSVR) to handle the classification problems with outliers or noises [37]. With these in mind, a new fuzzy rule-based inference model for intention understanding can be built, which integrates the advantages of fuzzy model and SVR. As a result, we consider applying our proposed Two-Layer Fuzzy SVR fused with TS fuzzy model (TLFSVR-TS) for people's intentions such as intention to drink at a bar.

The TLFSVR consists of two layers, including local and global learning layer. In the local learning layer, data classification is realized by fuzzy c-means (FCM), and SVR is used for data learning, in such a way that the training data is classified into multiple subsets, so that the conventional one SVR is updated to multiple SVRs. In the global learning layer, fuzzy weighted average algorithm is used to generate the intention. The TLFSVR-TS is the updated model from TLFSVR, and the biggest improvement is that the TS model-based fuzzy inference is substituted for the fuzzy weighted average algorithm, where intention is generated by fuzzy rules according to human identification information.

The main idea of TLFSVR-TS is that fuzzy SVR is used to classify input data according to three components of human properties, i.e., gender, province, and age, and the TS model-based fuzzy inference is used to generate the intention based on identification information. Specifically, at first, emotion of facial expression are recognized using candide3 based feature point matching method. After dynamic emotion recognition and considering human different identification information, emotional information are classified into different genders/provinces/ages based on fuzzy c-means. And then, based on different clusters of identification information and emotion information, deep understanding of human emotion is learned via multiple fuzzy support vector regression. Finally, human intention is generated by using TS model-based fuzzy inference, where fuzzy rules and membership functions are designed according to human information.

With TS model-based fuzzy inference, the TLFSVR-TS provides model modification by considering about the priori knowledge inferred from human personal preference, which can greatly reduce the uncertainty of communication environment in human-robot interaction, such as different people. In the case of various people with different genders/provinces/ages, it is very obvious that preferences (priori knowledge) of men and women are not the same. For example, when men feel sad or anger, most of them will strong desire for alcoholic drink, however, most of women will prefer non-alcoholic drink that when they have the feeling. Preference is different and thus affects people's intention to purchase products. Moreover, different regression models in multiple SVRs are corresponding to different clusters (classified by fuzzy c-means), which is better than only one SVR covers all situations in different genders/provinces/ages, and guarantee the local learning ability.

Identification information of human (i.e., gender, province, and age) has an important influence on emotional intention understanding according to characteristic analysis in our previous research [35], the proposal considers the effect of human identification information and emotional intention. In human-robot interaction, all the robots are mainly divided into two categories, including information robots and task robots. Information robot collects personal information including gender, province, and age, as well as what kind of drinks chosen in a different emotion. Task robot executes behaviors for adaptation. Structure of the proposal is shown in Fig. 10.1, which mainly consists of dynamic emotion recognition and emotional intention understanding.

Preliminary application experiments are developed to validate the proposal, where dynamic emotion recognition and understanding are established in the developing emotional social robot system (ESRS). In the experiments, according to a scenario

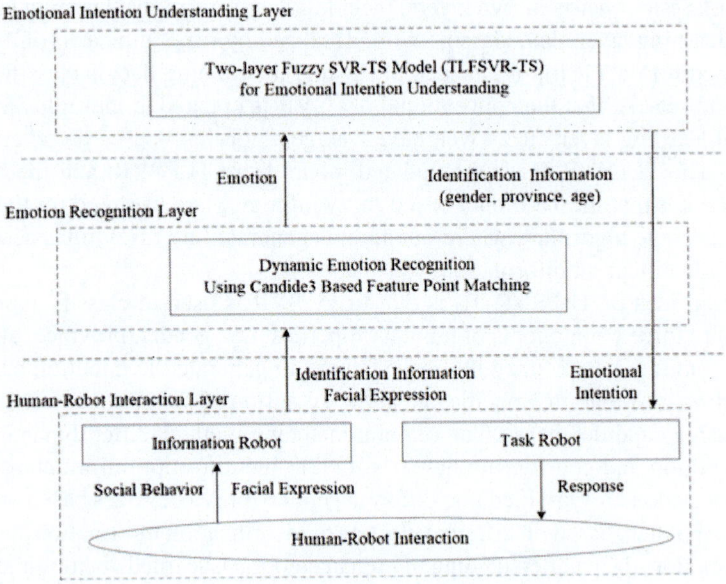

Fig. 10.1 Structure of dynamic emotion recognition and understanding

of "drinking at bar", two mobile robots are used as waiters and thirty postgraduates are invited as customers.

10.2 Dynamic Emotion Recognition Using Candide3-Based Feature Point Matching

The dynamic emotion recognition is mainly based on candide3 model, as shown in Fig. 10.2, which consists of feature points (marked in red point) and triangles (surrounded by yellow lines) of facial expression. In the candide3 model, facial action units (AU, consist of several triangles and each triangle is represented by feature points) are the basic element of the changes of the facial expression. Combination of facial AU is used to express the seven basic emotions, including happiness, anger, sadness, disgust, surprise, fear, and neutral. In the experiment, we find out six facial AU, including upper lip raiser (AU0), jaw lowerer (AU1), lip stretcher (AU2), brow lowerer (AU3), lip corner depressor (AU4), and outer brow raiser (AU5). Therefore, we analyze the changes of facial expressions by tracking facial AU, and produce the facial expression real-time. The extracted motion features of facial expressions can be represented as,

$$E_i = (AU_0, AU_1, \ldots, AU_m) \tag{10.1}$$

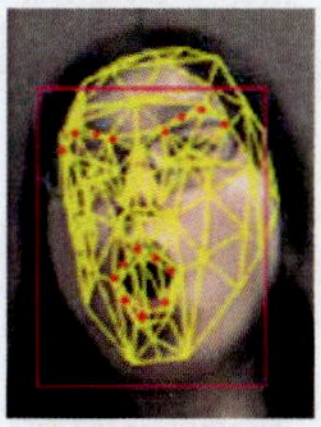

Fig. 10.2 Feature points and facial action units (AU) of candide3 model

where E_i means seven basic emotions and $AU_m \in (-1, 1)$ is correlation coefficient between real-time feature points and given feature points (recording in the database from volunteers) to express performance of each facial action units. For example, in AU0, $AU = 1$ means showing teeth fully and $AU = -1$ means maximal possible pushed down lip; in AU1, $AU = 1$ means fully open and $AU = -1$ means closed; in AU2, $AU = 1$ means fully stretched and $AU = -1$ means fully rounded; in AU3, $AU = 1$ means raised almost all the way and $AU = -1$ means fully lowered; in AU4, $AU = 1$ means very happy smile and $AU = -1$ means very sad frown; in AU5, $AU = 1$ means fully lowered as a very sad face and $AU = -1$ means raised as in an expression of deep surprise.

Algorithm 1 shows the candide3 based dynamic feature point matching algorithm, in which the emotion recognition results are obtained by the fuzzy rules in Table 10.1. According to the facial expression data collected from thirty volunteers in our lab, after analysis of comparisons and tests, we found that each of the seven emotion corresponding to different AU. For example, when the expression is obtained by surprise, the $AU_1 > 0$ is obvious, even to the extent that it appears $AU_1 > 0.1$, which is different from other expression, in such a way that fuzzy rules of emotion recognition are obtained in Table 10.1. In addition, we adopt sliding window mechanism. In real-time emotion recognition, it takes a few seconds for people to express one emotion, and the emotion is needed to keep few seconds for recognition. Considering about the consuming time of emotion expression and tracking the variation of the emotion in a short period, after repeated trials and tests, sliding window is set as 20 (acquisition time of AU is 0.5 s, and 20 sliding windows for recognition cost 10 s). If the same AU appears more, then it belongs to a certain kind of emotion.

Table 10.1 Fuzzy rules of emotion recognition

Rule	If	Then
R_1	$AU_3 > 0.1$, $AU_5 > 0.05$	Anger
R_2	$AU_5 > 0.05$, $AU_2 < -0.2$	Fear
R_3	$AU_1 > 0.1$, $AU_3 < 0.0$	Surprise
R_4	$AU_2 - AU_4 > 0.1$, $AU_4 < -0.3$	Happiness
R_5	$AU_3 > 0.1$, $AU_1 < 0$	Sadness
R_6	$AU_1 < 0$, AU_2 <-0.3, $AU_0 > 0.1$	Disgust
R_7	$-0.1 < AU_m < 0.1$	Neutral

Algorithm 1. Candide3 based dynamic feature point matching algorithm

1. Initialization.
   Input: $AU_m, m = 0, 1, 2, \ldots, 5$.
   Output: $E_i, i = 1, 2, 3, \ldots, 7$.
   where $AU_m \in (-1, 1)$ is the correlation coefficient to express the performance of each facial action units and E_i represents 7 basic emotions.
2. Termination Check.
   (a) If all of the AU_m match one of the fuzzy rules, then it is selected and E_i will be the output solution.
   (b) Else if no fuzzy rules match all of the AU_m, then fuzzy rules that matches most of the AU_m is selected and E_i will be the output solution.
   (c) Else goto the 3.
3. For $i = 1, \ldots n$, Do
   (a) Update AU_m according to correlation coefficient between real-time and given feature points.
   (b) Obtain facial expression E_i, according to the Table 3.2.
   (c) If the E_i appears more in a sliding window with 20 intervals, then the E_i will be the output solution.
4. End For.
5. Update E_i.
6. Goto 2.

10.3 Two-Layer Fuzzy Support Vector Regression for Emotional Intention Understanding

SVR is being more and more popular in the application of humans and robots interaction, at the same time, it faces some challenges, the important one is that the local modeling performance is far from satisfaction, in other word, only one regression hyperplane cannot cover most training samples. In the case of people with different ages, to describe different situations such as own hobbies, different regression models could be better than only one of it. Local learning algorithm aims at locally adjusting the model adaptive capacity according to the features of the training data subset [38]. TLFSVR is proposed based on the idea of local learning to solve the evenly distributed problem. TLFSVR consist of two-layer as shown in Fig. 10.3, which includes local and global learning layer. In the local learning layer, fuzzy c-means is used for the data classification, and SVR is used for data learning. In the global learning layer, the intention is obtained based on fuzzy weighted average algorithm.

At first, based on the fuzzy c-means [39, 40], the training data are classified into several subsets, and it aims to minimize the objective function as follow,

$$f(U, V) = \sum_{i=1}^{l} \sum_{k=1}^{C} (\mu_{ik})^m \|X_i - V_k\|^2 \tag{10.2}$$

where $l \in N$ denotes training data number, $C \in [2, l)$ denotes clusters number, $m \in [1, +\infty)$ denotes a weighting exponent, $\mu_{ik} \in [0, 1]$ denotes the membership grade of training data X_i in the cluster k, V_k denotes the center of the cluster k.

Approximate optimization of (10.7) is via iteration, with the updating of membership and the cluster centers V_k by

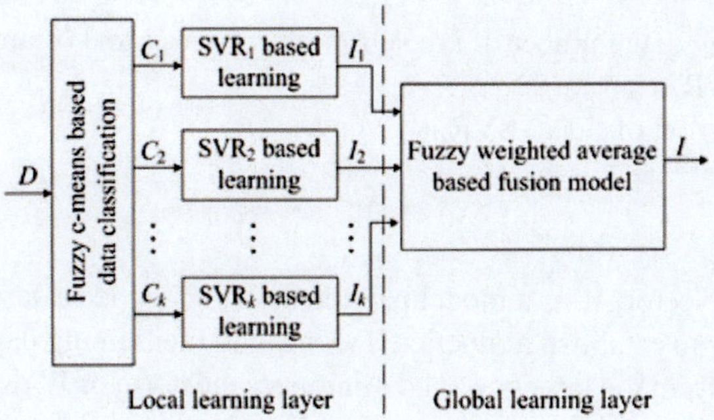

Fig. 10.3 Two-layer fuzzy support vector regression (TLFSVR)

$$\mu_{ik} = \left(\sum_{k=1}^{C} \left(\frac{d_{ij}}{d_{ik}} \right)^{\frac{2}{(m-1)}} \right)^{-1}, 1 \leqslant i \leqslant l, 1 \leqslant j \leqslant C \tag{10.3}$$

$$V_k = \frac{\sum_{i=1}^{l} (\mu_{ik})^m X_i}{\sum_{i=1}^{l} (\mu_{ik})^m}, 1 \leqslant j \leqslant C \tag{10.4}$$

where $d_{ik} = \|X_i - V_k\|$.

While the termination criterion is satisfied, i.e., $\|X_i - V_k\| \leqslant \epsilon$, the iteration will stop, where $V = (v_1, v_2, \ldots, v_C)$, and ϵ is sensitivity threshold.

The spread width is obtained by

$$\delta_k^o = \sqrt{\frac{\sum_{i=1}^{l} (\mu_{ik})^m \left\| x_i^o - v_i^o \right\|^2}{\sum_{i=1}^{l} (\mu_{ik})^m}}, 1 \leqslant k \leqslant C, 1 \leqslant o \leqslant N. \tag{10.5}$$

Based on the spread width and cluster centers, training data is separated into several subsets as (10.11),

$$D_k = \left\{ (X_i, y_i) \| v_k^o - \eta \delta_k^o \leqslant x_i^o \leqslant v_k^o + \eta \delta_k^o \right\}, \quad 1 \leqslant i \leqslant l, 1 \leqslant o \leqslant N, 1 \leqslant k \leqslant C \tag{10.6}$$

where η denotes a constant, which is to control the overlap region of the training subsets, and it will increase as the size of training subsets increase.

Secondly, each cluster's regression function is calculated as shown below,

$$SVR_k = \sum_{i=1}^{l_k} \left(\alpha_{i,k}^* - \alpha_{i,k} \right) k\left(X_i, X \right) + b_k, X \in D_k \tag{10.7}$$

where l_k is training data number of kth subset, the $\alpha^*_{i,k}$, $\alpha_{i,k}$ and b_k are calculated via the training of SVR.

The basic function of linear SVR is given as

$$f(X) = \langle W, X\rangle + b \tag{10.8}$$

where X is input vector, W is a model parameter, and b is the scalar bias.

SVR is aiming to establish a function $f(X)$ for all the training data, which has at most ε deviation from the targets y_i, and minimizes the norm of W as far as possible. Convex optimization problem can be used to describe this problem and written as

$$\begin{array}{ll} \min\limits_{W,b,\xi_i,\xi_i^*} & \frac{1}{2}\|W\|^2 + \gamma \sum\limits_{i=1}^{l} (\xi_i + \xi_i^*) \\ \text{subject to} & y_i - \langle W, X_i\rangle - b \leqslant \varepsilon + \xi_i \\ & \langle W, X_i\rangle + b - y_i \leqslant \varepsilon + \xi_i^* \\ & \xi_i, \xi_i^* \geqslant 0, i = 1, 2, \ldots, l \end{array} \tag{10.9}$$

where γ is a scalar regularization parameter, and slack variables ξ_i and ξ_i^* are introduced to handle the otherwise infeasible constrains.

Optimization of hyperplane in (10.3) is a QP solver, Lagrangian is constructed and transformed into the dual to solve the problem as

$$\begin{array}{ll} \max\limits_{\alpha,\alpha^*} & -\frac{1}{2}\sum\limits_{i,j=1}^{l} (\alpha_i - \alpha_i^*)\left(\alpha_j - \alpha_j^*\right)\langle X_i, X_j\rangle \\ & -\varepsilon \sum\limits_{i=1}^{l} (\alpha_i + \alpha_i^*) + \sum\limits_{j=1}^{l} y_i (\alpha_i - \alpha_i^*) \\ \text{subject to} \sum\limits_{i=1}^{l} (\alpha_i - \alpha_i^*) = 0 & and \quad \alpha_i, \alpha_i^* \in [0, \gamma] \end{array} \tag{10.10}$$

where α_i and α_i^* mean Lagrange multipliers and dot product is kernel function.

Finally, by using the fusion model of fuzzy weighted average algorithm [41], output of global learning is constructed based on the multi-SVRs and membership functions. The membership function is defined as,

$$A_k(X_i) = \max\left(\min\left(\frac{x_i^o - (v_k^o - \eta\delta_i^o)}{v_k^o - (v_k^o - \eta\delta_k^o)}, \frac{(v_k^o + \eta\delta_k^o) - x_i^o}{(v_k^o + \eta\delta_k^o) - v_k^o}\right), 0\right) \tag{10.11}$$

Then, output of the global learning is calculated as follows,

$$I_i(X_i) = \frac{\sum\limits_{k=1}^{C} A_k(X_i) SVR_k(X_i)}{\sum\limits_{k=1}^{C} A_k(X_i)}, 1 \leqslant i \leqslant l, 1 \leqslant k \leqslant C. \tag{10.12}$$

10.4 Two-Layer Fuzzy Support Vector Regression Takagi-Sugeno Model for Emotional Intention Understanding

To describe a complicated nonlinear system, the TS fuzzy model is used to decompose input sets into several subsets, in such a way that each subset can be expressed via a simple linear regression mode. The typical TS fuzzy rule is shown as follow,

R^i: If $z_1(t)$ is F_i^1 and if $z_2(t)$ is F_i^2 and... and if $z_n(t)$ is F_i^n, then $\dot{x}(t) = Ax(t)$.

The nonlinear model is given in the form,

$$\dot{x}(t) = f(x(t)) \tag{10.13}$$

where $f(\cdot)$ means nonlinear function.

Deriving from TS fuzzy model, a model is expressed as,

$$\dot{x}(t) = \sum_{i=1}^{R} h_i(z(t))A_i x(t) = A_z x(t) \tag{10.14}$$

where $x \epsilon \mathbb{R}^n$ denotes the state vector, $h_i(z)$ denotes the membership functions, $z(t)$ denotes the premise variable with bounds.

The membership function is calculated by

$$h_i(z) = \frac{\mu_i(z)}{\sum_{i=1}^{R} \mu_i(z)} \tag{10.15}$$

where $\mu_i(z) = \prod_{j=1}^{n} F_i{}^j(z_j)$. Note that $h_i(\cdot)$ satisfies and $h_i(\cdot) \geq 0$.

If the input vector $X = [x_1, x_2, \ldots x_N]$ is given, then the output can be obtained via,

$$y_i = \alpha_i{}^0 + \alpha_i{}^1 x_1 + \cdots + \alpha_i{}^N x_N \tag{10.16}$$

$$\hat{y} = \frac{\sum_{i=1}^{R} h_i y_i}{\sum_{i=1}^{R} h_i} \tag{10.17}$$

where α_i^j is the consequent parameter of the ith output y_i.

The TLFSVR-TS is the updated model from TLFSVR, in which the TS model-based fuzzy inference is substituted for the fuzzy weighted average algorithm as shown in Fig. 10.4. In TLFSVR, input D includes emotion, gender, province, and age, while output is the drinking intention. In the local learning layer, emotional information are classified into different clusters C_i of genders/provinces/ages based on fuzzy c-means, and then deep understanding of human emotion I_i is learned via multiple fuzzy support vector regression. In the global learning layer, human intention I is generated by using TS model-based fuzzy inference, where fuzzy rules and membership functions are designed according to human information.

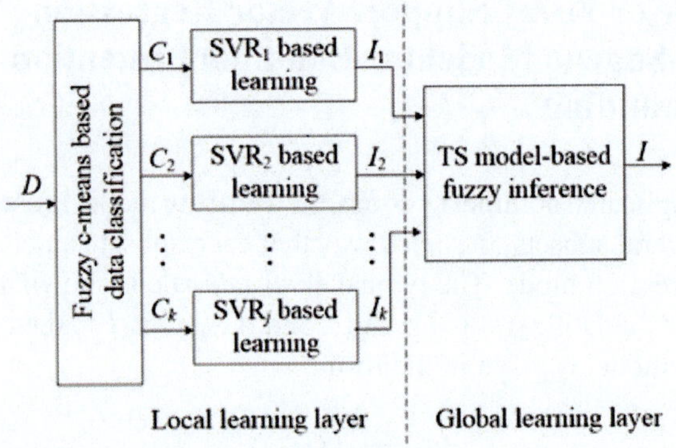

Fig. 10.4 Two-layer fuzzy SVR-TS model (TLFSVR-TS)

Algorithm 2 shows the details of Two-layer Fuzzy SVR-TS Algorithm. Thirty volunteers' intenstions to drink in the home environment have been collected for building the fuzzy rules, where volunteers write down the intensity value of intention (II_i) from 1 to 5 (i.e., very week, week, neutral, strong, very strong) according to seven basic emotions (E_i). From the collected data, we can find that women's and men's drinking intention is different in different emotion. In such a way that the TLFSVR-TS model from (10.15)–(10.18) are revised as follow,

R^i: If $E = E_i$, $I = I_i$, $II = II_i$, then

$$y_i = \begin{cases} I_{TLFSVR} \ \& \ h_{TLFSVR} = II_{TLFSVR} / \sum_{j=1}^{R_n} II_j, & I_i = I_{TLFSVR} \\ I_i \ \& \ h_i = II_i / \sum_{j=1}^{R_n} II_j, & I_i \neq I_{TLFSVR} \end{cases} \tag{10.18}$$

where i is the number of fuzzy rules, from 1 to 56 (= 7 emotions $\times$ 8 intentions), and each emotion corresponds to 8 intentions, thus $R_n = 8$.

Algorithm 2. Two-layer Fuzzy SVR-TS Algorithm

---

1. Initialization.
   Input: X, C, l, V.
   Output: y.
   where X is the training data including emotion, gender, province, and age, $l \in N$ is the number of training data, C is the number of clusters, V is the center of cluster and $V = (v_1, v_2, \ldots v_C)$, y is the drinking intention.
2. Termination Check.
   (a) If $\| X_i - V_k \| \geq \epsilon$, then the iteration will stop, where X and V are normalized, and $\epsilon = 1$.
   (b) Else goto 3.
3. For $i = 1, \ldots l$, Do
   (a) Update the membership and the cluster centers, according to (10.8) and (10.9), and the output $y(X_i)$.
   (b) Calculate the spread width, according to (10.10), then divide training data into several subsets.
   (c) Take gender, province, age into account, the k_{th} regression function is calculated according to (10.12) and optimized via (10.4).
   (d) Obtain the fuzzy weighted of the k_{th} SVR, according to (10.13).
   (f) Get the output $y(X_i)$ of the proposed TLFSVR-TS according to (10.18) and (10.19).
4. End For.
5. Update $y(X_i)$.
6. Goto 2.

Then the output of the proposed TLFSVR-TS can be calculated as

$$\hat{y} = \frac{\sum_{i=1}^{R_n} h_i y_i}{\sum_{i=1}^{R_n} h_i} \tag{10.19}$$

10.5 Experiments

Two-layer Fuzzy SVR-TS fuzzy model (TLFSVR-TS) is proposed for emotional intention understanding in human-robot interaction, where TS model-based fuzzy inference is substituted for the fuzzy weighted average algorithm, and intention is generated by fuzzy rules according to human identification information. In this part, the data settings and experimental environment settings are explained. The simulation experiment is carried out and the experimental results are analyzed.

10.5.1 Experimental Environment

The proposed intention understanding model is verified by using the developed emotional social robot system (ESRS) in our lab, where two emotional robots (information robot and deep cognitive robot) and thirty customers (volunteers from postgraduates) experience the scenario of "drinking at a bar", as shown in Fig. 10.5. ESRS consist of information sensing and transfer layer and intelligent computation layer, as shown in Fig. 10.6.

10.5.2 Self-Built Data

30 volunteers aged 18–28 are invited to build up the experimental data; they are postgraduates of our lab or undergraduate students. All of them come from China from different provinces, and the influence of province is also taken into account.

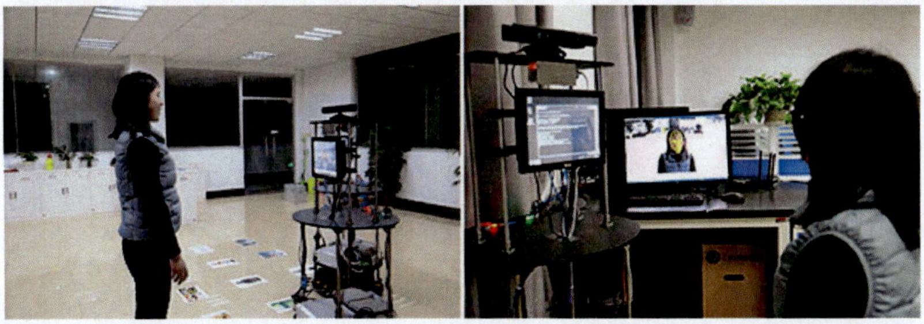

Fig. 10.5 Experimental environment

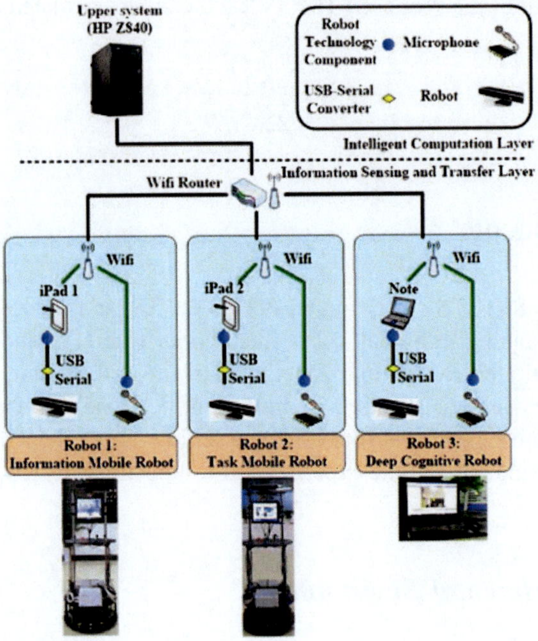

Fig. 10.6 Network of ESRS

Each volunteer expresses seven primary emotions, i.e., happiness, anger, sadness, disgust, surprise, fear, and neutral, and human emotional intention includes the 8 kinds such as 1-wine, 2-beer, 3-sake, 4-sour, 5-shochu, 6-whisky, 7-non-alcoholic drink, and 8-others (other drinks or food). The desire for each drink is divided into 5 levels, i.e.,very weak (1), weak (2), neutral (3), strong (4), and very strong (5). Volunteers can choose 1 to 5 in the questionnaire for express the desire for each drink in different emotion. For information mobile robot, candide3 based dynamic feature point matching method is used to recognize the dynamic emotion; for deep cognitive robot, human emotional intention is understood by using proposed TLFSVR-TS model. Data samples of one volunteer are collected as shown in Fig. 10.7. In fact, 420 groups of data are collected in total, where half of data was taken for training and other half for testing.

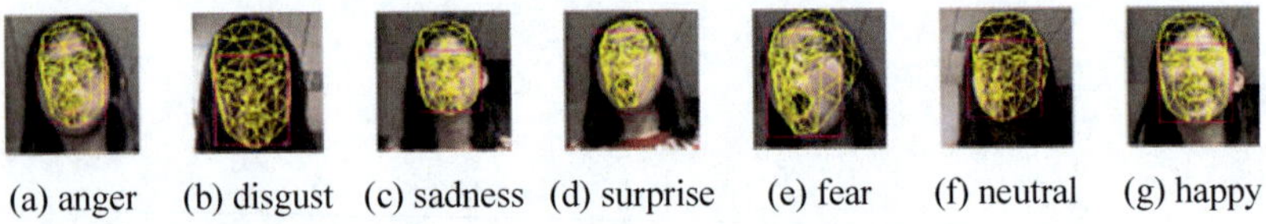

Fig. 10.7 Experimental volunteer

10.5.3 *Experiments on Dynamic Emotion Recognition and Understanding*

Dynamic emotion recognition and emotion understanding are both done in the preliminary application. According to the real-time dynamic facial expression of humans, 210 dynamic emotions of 30 volunteers are recognized by using the feature point matching and the introduced Candide3 based dynamic feature points matching. Experiment results are shown in Table 10.2, and confusion matrices of emotion recognition are shown in Tables 10.3 and 10.4.

According to analyses of results of average recognition rate and confusion matrices, the average recognition rate is 80.48% by using the proposal, which is 10.48% higher than that of feature point matching. Due to the sliding window mechanism, which is used to select the 20 sets of the dynamic emotion for repeated confirming the emotion, the recognition rate is high than the other method. Moreover, the proposal takes into account correlations of feature points in each facial action units (AU), which can find out the variation of main feature points in facial expression,

Table 10.2 Comparison of dynamic emotion recognition experimental results

Algorithm	Average recognition rate (%)
Feature point matching	70
Candide3 based dynamic feature point matching	80.48

Table 10.3 Confusion matrices by using feature point matching

Emotion	Sadness	Angry	Surprise	Disgust	Fear	Neutral	Happiness
Sadness (%)	43.34	0	0	33.33	0	23.33	0
Angry (%)	0	43.34	0	33.33	23.33	0	0
Surprise (%)	0	6.67	86.66	6.67	0	0	0
Disgust (%)	0	20.00	0	66.67	13.33	0	0
Fear (%)	0	20.00	0	16.67	63.33	0	0
Neutral (%)	0	0	0	0	0	100	0
Happiness (%)	0	0	0	0	0	13.33	86.67

Table 10.4 Confusion matrices by using candide3 based dynamic feature point matching

Emotion	Sadness	Angry	Surprise	Disgust	Fear	Neutral	Happiness
Sadness (%)	66.67	0	0	23.33	0	10.00	0
Angry (%)	0	63.34	0	23.33	13.33	0	0
Surprise (%)	0	3.33	93.34	3.33	0	0	0
Disgust (%)	0	13.33	0	76.67	10.00	0	0
Fear (%)	0	10.00	0	10.00	80.00	0	0
Neutral (%)	0	0	0	0	0	100	0
Happiness (%)	0	0	0	0	0	16.67	83.33

in such a way that improving emotion recognition rate. Since emotion expression is varied person to person, the proposal can handle such problem in two ways. One is that database of feature points corresponding to each emotion are collected from volunteers (consist of male and female). In the case of stranger, we can match the real-time data to the database. Because different genders have their own feature, the matching can be divided into two cluster, i.e., male and female, and in such a way that the recognition rate can be improved. The other is that the fuzzy rules for emotion recognition are established. In the case of no full matching of the feature points, fuzzy rules that match most of the feature points will be selected, which guarantees the recognition rate to some extent.

To show the efficiency of the proposed emotion understanding model, comparative experiments are carried out among the TLFSVR-TS, the TLFSVR [35], KFCM-FSVR [37], and the SVR, which aims to understand the intention and discuss the impact of gender/province/age on the emotional intention understanding. Furthermore, to establish TS fuzzy model, the chapter analyzes thirty volunteers' intention to drink in the home environment. According to data analysis, we found that preferences of men and women are different in the home environment. Moreover, it is not difficult to find that most people would like to drink beer and shochu when they feel displeasure. But in turn, they prefer non-alcoholic drink or foods. This maybe relate to the Chinese eating habits. As a result, this can serve as reference of drinking habits in the home environment.

In the experiments, inputs are emotion, gender, province, and age, output is the intention. Clusters for FCM and membership function of TS fuzzy model are consider for developing the experiments. According to the gender (i.e., female and male), the number of clusters C for FCM is set as 2; considering people from different provinces in China, i.e., Hubei province (location of our lab) and non-Hubei province, the number of clusters C for the FCM is set as 2; since postgraduates (volunteers) can be divided as senior, junior, and freshman according to age, the number of clusters C is set as 3.

Based on the experimental experience in our previous study [35], the value of the parameters of clustering methods are obtained by trial and error. For parameters setting of FCM, the overlap parameter η can be chosen as 2.5, 3.5, 4, and 5 according to input dimension with 1, 2, 4, and 8, respectively. With this in mind, since there are 4 inputs in the experiments, the η and the sensitivity threshold ϵ are set to 3 and 1, respectively. In the SVR, width σt of Gaussian kernel is recommended as $(0.1 - 0.5) \times$ (rang of the inputs), and according to the inputs in the experiments are all normalized, σ can be select as 0.5, and the scalar regularization parameter γ is set as 200.

Figures 10.8, 10.9, 10.10, 10.11, 10.12, 10.13, 10.14, 10.15 and 10.16 show the performance of the proposed TLFSVR-TS, TLFSVR, KFCM-FSVR and SVR with $C = 2/2/3$ in different membership functions (MF), respectively, and the results analysis are showed in Table 10.5. In the experiments, standard deviation (SD), correlation coefficient (CC), understanding accuracy (UA) are used as evaluation indexes. According to experimental results, in the case of MAX membership function, TLFSVR-TS model obtains understanding accuracy of 76.67, 76.19, 75.71%,

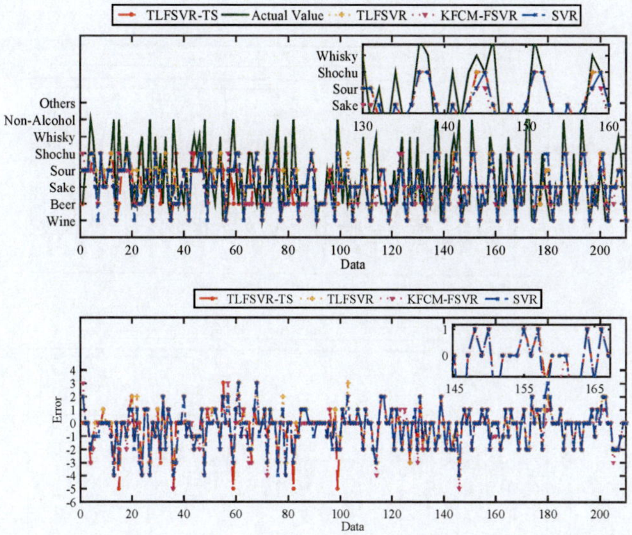

Fig. 10.8 Emotion understanding using MAX membership function with $C = 2$ (Gender)

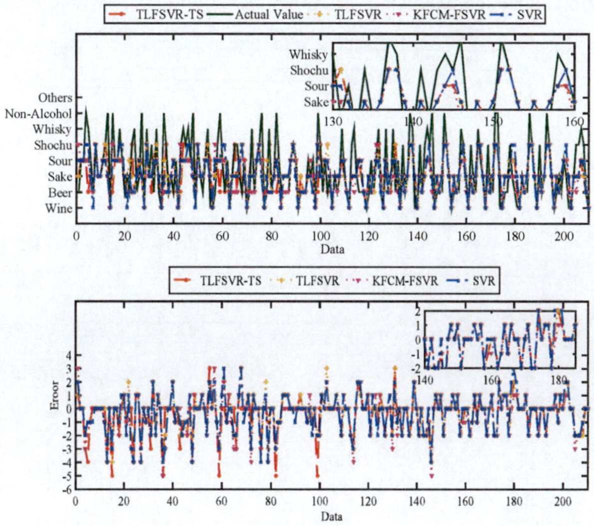

Fig. 10.9 Emotion understanding using MAX membership function with $C = 2$ (Province)

while $C = 2/2/3$, as shown in Figs. 10.8, 10.9 and 10.10. In the case of MEAN membership function, TLFSVR-TS model obtains understanding accuracy of 76.19, 75.71, 75.24%, while $C = 2/2/3$, as shown in Fig. 10.11, 10.12 and 10.13. In the case of MEDIAN-MAX membership function, TLFSVR-TS model obtains understanding accuracy of 77.67, 76.19, 75.71%, while $C = 2/2/3$, as shown in Fig. 10.14, 10.15 and 10.16. The results demonstrate that the proposed TLFSVR-TS model produces high accuracy than those of TLFSVR, KFCM-FSVR, and SVR.

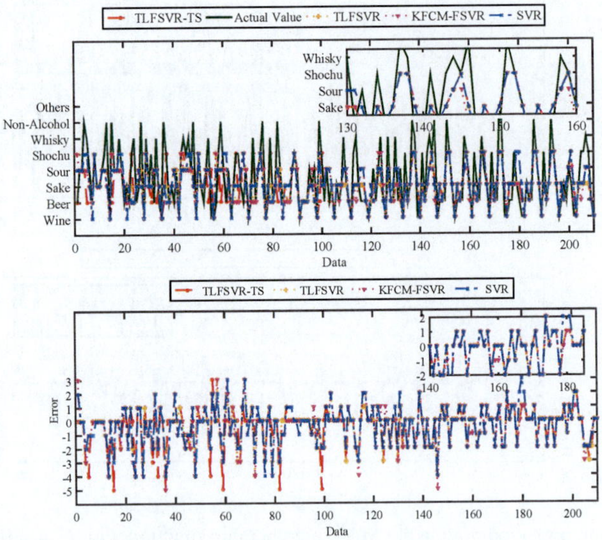

Fig. 10.10 Emotion understanding using MAX membership function with $C = 3$ (Age)

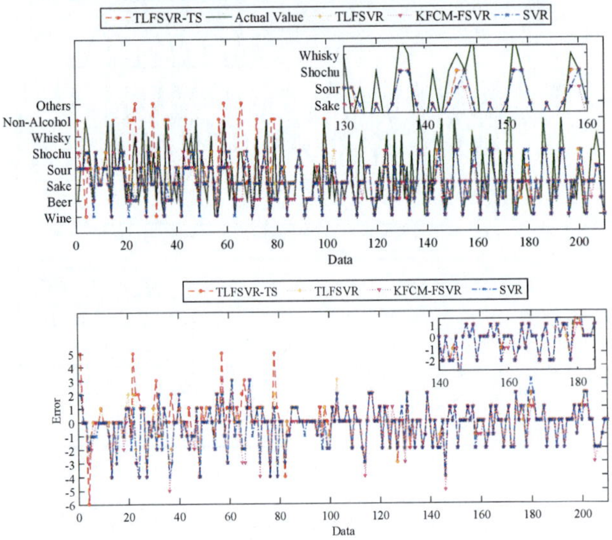

Fig. 10.11 Emotion understanding using MEAN membership function with $C = 2$ (Gender)

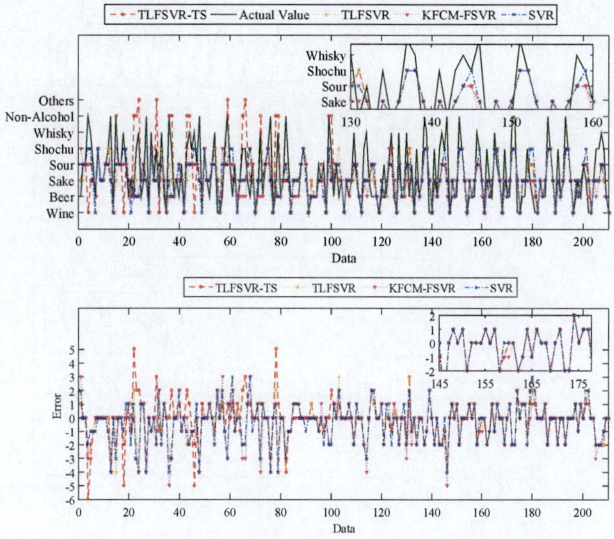

Fig. 10.12 Emotion understanding using MEAN membership function with $C = 2$ (Province)

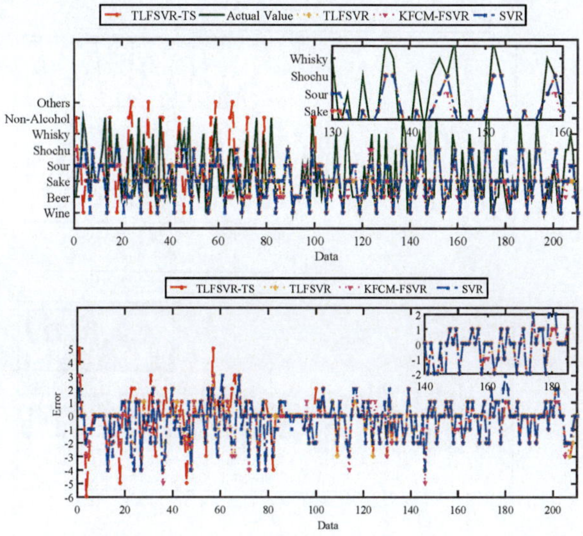

Fig. 10.13 Emotion understanding using MEAN membership function with $C = 3$ (Age)

According to results of the experiments, the proposed emotion understanding model yields higher accuracy than that of the TLFSVR, the KFCM-FSVR, and the SVR, leading to higher correlation coefficient and smaller squared error between the actual values and the output values. Higher correlation and smaller squared error means that our proposal can better reflect the actual situation. At the same time, by changing the number of clusters C according to the identification information (i.e., gender, province, and age) and membership function (MF) based on data collection

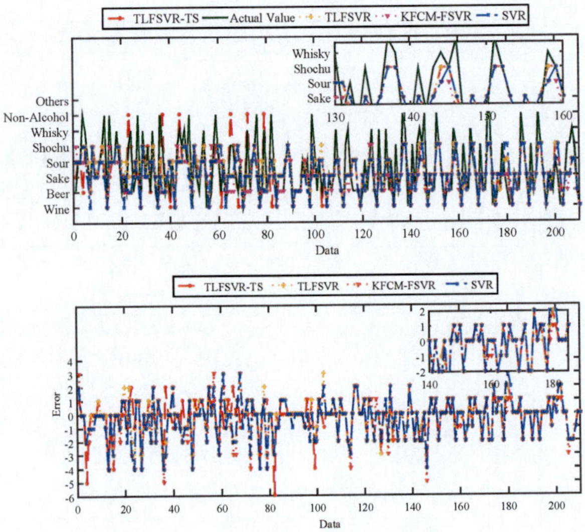

Fig. 10.14 Emotion understanding using MEDIAN-MAX membership function with $C = 2$ (Gender)

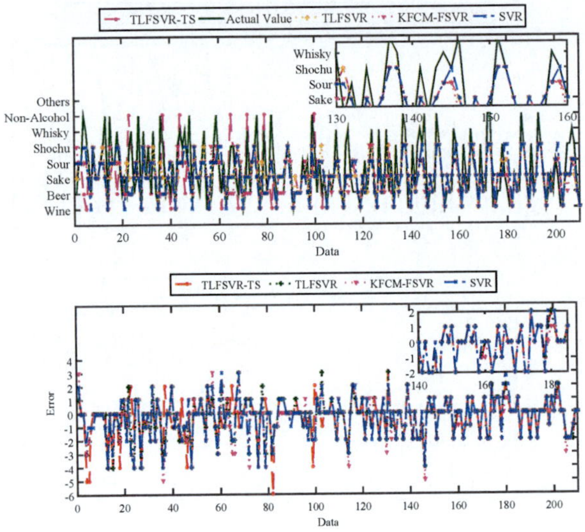

Fig. 10.15 Emotion understanding using MEDIAN-MAX membership function with $C = 2$ (Province)

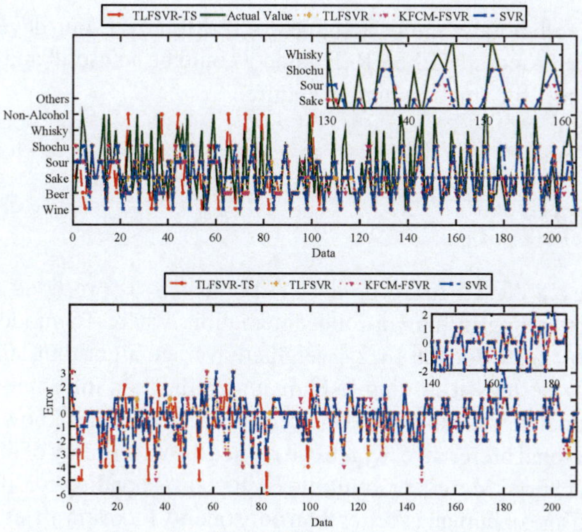

Fig. 10.16 Emotion understanding using MEDIAN-MAX membership function with $C = 3$ (Age)

Table 10.5 Comparison of emotion understanding

Index	MF	TLFSVR-TS [16]			TLFSVR			KFCM-FSVR [40]			SVR
		Gender	Province	Age	Gender	Province	Age	Gender	Province	Age	
		$C = 2$	$C = 2$	$C = 3$	$C = 2$	$C = 2$	$C = 3$	$C = 2$	$C = 2$	$C = 3$	
SD	MAX	1.4130	1.4220	1.4511	1.4042	1.4249	1.4798	1.3971	1.4034	1.4739	1.4667
	MEAN	1.5123	1.5159	1.5377							
	MEDIAN-MAX	1.3974	1.4630	1.5264							
CC	MAX	0.6615	0.6561	0.6378	0.6710	0.6698	0.6580	0.6680	0.6560	0.6190	0.6278
	MEAN	0.6414	0.6361	0.6204							
	MEDIAN-MAX	0.6751	0.6379	0.6008							
UA (%)	MAX	76.67	76.19	75.71	70.00	70.00	69.05	70.48	69.05	68.10	67.62
	MEAN	76.19	75.71	75.24							
	MEDIAN-MAX	77.62	76.19	75.71							

of volunteers, appropriate C and MF can be found to enhance the understanding accuracy to a certain extent. Additionally, supplementary information such as identification information, are applied to minimize the influence of missing or false emotion recognition, and in such a way that ensure the accuracy of emotion understanding.

For further research in human-robot interaction, deep level information understanding is being more and more popular, how robots adapt their behaviors to deep level information such as intention, but not only emotion and atmosphere [42, 43], would be an interesting research topic. With the rapid progress of the affective robots, they have greater intelligence, for example, at home environment, Pepper [44] learns together with children; based on people's actions understanding, robot plays as sociable partners in collaborative joint activities [45]. Moreover, our developing ESRS also indicates the proposed TLFSVR-TS model could be an useful and practical way to emotional robots for emotion understanding.

10.6 Summary

Two-layer Fuzzy SVR-TS fuzzy model (TLFSVR-TS) is proposed for emotional intention understanding in human-robot interaction, where TS model-based fuzzy inference is substituted for the fuzzy weighted average algorithm, and intention is generated by fuzzy rules according to human identification information. TLFSVR-TS provides model modification by considering about the priori knowledge inferred from human personal preference, which can reduce the uncertainty of different people in the communication. Moreover, multiple SVRs corresponding to different gender, providence, and age of human is better than only one SVR covers all situations, which guarantees the local learning ability. Experimental results show that the proposal receives higher accuracy than that of the TLFSVR, the KFCM-FSVR, and the SVR with higher correlation coefficient and smaller squared.

Deep level information understanding is getting more popular in human-robot interaction. Robots behavior adaptation to human intention would be an interesting research topics. The preliminary application experiments via our ESRS, gives the strong evidence that the proposed TLFSVR-TS model is suitable to understand human emotional intention for emotional robots.

References

1. A. Tawari, M.M. Trivedi, Face expression recognition by cross modal data association. IEEE Trans. Multimed. **15**(7), 1543–1552 (2013)
2. P. Melin, O. Mendoza, O. Castillo, Face recognition with an improved interval type2 fuzzy logic sugeno integral and modular neural networks. IEEE Trans. Syst. Man Cybern. Part A Syst. Hum. **41**(5), 1001–1012 (2011)
3. W. Di, L. Zhang, D. Zhang et al., Studies on hperspectral face recognition in visible spectrum with feature band selection. IEEE Trans. Syst. Man Cybern. Part A Syst. Hum. **40**(6), 1354–1361 (2010)
4. M. J. Salvador, S. Silver, M.H. Mahoor, An emotion recognition comparative study of autistic and typically-developing children using the zeno robot, in *Proceedings of IEEE International Conference on Robotics and Automation* (Washington, USA, 2015), pp. 6128–6133
5. A. Halder, A. Konar, R. Mandal et al., General and interval type-2 fuzzy face-space approach to emotion recognition. IEEE Trans. Syst., Man, Cybern. Syst. **43**(3), 587–605 (2013)
6. S. Walter, J. Kim, D. Hrabal et al., Transsituational individual-specific biopsychological classification of emotions. IEEE Trans. Syst., Man, Cybern. Syst. **43**(4), 988–995 (2013)
7. D. Wu, Fuzzy sets and systems in building closed-loop affective computing systems for Human-robot interaction: advances and new research directions, in *Proceedings of IEEE International Conference on Fuzzy Systems* (Niskayuna, USA, 2012), pp. 1–8

8. Y. Sha, Emotional intelligence and affective computing in uncertainty situation, in *Proceedings of 3th International Conference on Intelligent System Design and Engineering Applications(ISDEA)* (Hong Kong, China, 2013), pp. 705–708
9. F.J.M. Angel, A. Bonarini, Studying people's emotional responses to robot's movements in a small scene, in *Proceedings of IEEE International Symposium on Robot and Human Interactive Communication* (Edinburgh, England, 2014), pp. 417–422
10. L. Bi, O. Tsimhoni, Y. Liu, Using the support vector regression approach to model human performance. IEEE Trans. Syst. Man Cybern. Part A Syst. Hum. **41**(3), 410–417 (2011)
11. C. Jiang, Z. Ni, Y. Guo, H. He, Learning human-robot interaction for robot-assisted pedestrian flow optimization. IEEE Trans. Syst., Man, Cybern. Syst. (2017). https://doi.org/10.1109/TSMC.2017.2725300
12. W.H. Al-Arashi, H. Ibrahim, S.A. Suandi, Optimizing principal component analysis performance for face recognition using genetic algorithm. Neurocomputing **128**, 415–420 (2014)
13. C. Zhou, L. Wang, Q. Zhang et al., Face recognition based on PCA and logistic regression analysis. Optik - Int. J. Light. Electron Opt. **125**(20), 5916–5919 (2014)
14. Z. Wang, Q.Q. Ruan, G.Y. An, Facial expression recognition using sparse local Fisher discriminant analysis. Neurocomputing **174**, 756–766 (2016)
15. D. Smeets, P. Claes, J. Hermans et al., A comparative study of 3-D face recognition under expression variations. IEEE Trans. Syst., Man, Cybern., Part C **42**(5), 710–727 (2012)
16. E. Vezzetti, F. Marcolin, G. Fracastoro, 3D face recognition: an automatic strategy based on geometrical descriptors and landmarks. Robot. Auton. Syst. **62**(12), 1768–1776 (2014)
17. F. Sun, G.B. Huang, Q.M.J. Wu et al., Efficient and rapid machine learning algorithms for big data and dynamic varying systems. IEEE Trans. Syst., Man, Cybern.: Syst. **47**(10), 2625–2626 (2017)
18. C.K. Hsieh, S.H. Lai, Y.C. Chen, An optical flow-based approach to robust face recognition under expression variations. IEEE Trans. Image Process. **19**(1), 233–240 (2010)
19. J.N. Wang, R. Xiong, J. Chu, Facial feature points detecting based on Gaussian mixture models. Pattern Recognit. Lett. **53**, 62–68 (2015)
20. S. Asteriadis, N. Nikolaidis, I. Pitas, Facial feature detection using distance vector fields. Pattern Recogn. **42**(7), 1388–1398 (2009)
21. T.F. Cootes, G.J. Edwards, C.J. Taylor, Active appearance models. IEEE Trans. Pattern Anal. Mach. Intell. **23**(6), 681–685 (2001)
22. Y.M. Lui, J.R. Beveridge, L.D. Whitley, Adaptive appearance model and condensation algorithm for robust face tacking. IEEE Trans. Syst. Man Cybern. Part A Syst. Hum. **40**(3), 437–448 (2010)
23. I. Kotsia, I. Pitas, Facial expression recognition in image sequences using geometric deformation features and support vector machines. IEEE Trans. Image Process. **16**(1), 172–187 (2007)
24. J. Wang, D. Yang, W. Jiang et al., Semisupervised incremental support vector machine learning based on neighborhood kernel estimation. IEEE Trans. Syst., Man, Cybern.: Syst. **47**(10), 1–11 (2017)
25. W. Yang, X. Sun, Q. Liao, Cascaded elastically progressive model for accurate face alignment. IEEE Trans. Syst., Man, Cybern.: Syst. **47**(9), 2613–2621 (2017)
26. Q. Hu, S. Zhang, M. Yu et al., Short-term wind speed or power forecasting with heteroscedastic support vector regression. IEEE Trans. Sustain. Energy **7**(1), 241–249 (2016)
27. W.Y. Cheng, C.F. Juang, A fuzzy model with online incremental SVM and margin-selective gradient descent learning for classification problems. IEEE Trans. Fuzzy Syst. **22**(2), 324–337 (2014)
28. C.F. Juang, R.B. Huang, W.Y. Cheng, An interval type-2 fuzzy-neural network with support-vector regression for noisy regression problems. IEEE Trans. Fuzzy Syst. **18**(4), 686–699 (2010)
29. X. Zhao, Y. Yin, L. Zhang et al., Affine TS fuzzy model-based estimation and control of hindmars- rose neuronal model. IEEE Trans. Fuzzy Syst. **24**(1), 235–241 (2015)

30. S. Beyhan, Exact output regulation for nonlinear systems described by Takagi-Sugeno fuzzy models. IEEE Trans. Syst., Man, Cybern.: Syst. 1–9 (2017)
31. R. Robles, M. Bernal, Comments on exact output regulation for nonlinear systems described by Takagi-Sugeno fuzzy models. IEEE Trans. Fuzzy Syst. **23**(1), 230–233 (2015)
32. W.A. De Souza, M.C.M. Teixeira, R. Cardim et al., On switched regulator design of uncertain nonlinear systems using Takagi-Sugeno fuzzy models. IEEE Trans. Fuzzy Syst. **22**(6), 1720–1727 (2014)
33. N.J. Cheung, X.M. Ding, H.B. Shen, OptiFel: a convergent heterogeneous particle swarm optimization algorithm for Takagi-Sugeno fuzzy modeling. IEEE Trans. Fuzzy Syst. **22**(4), 919–933 (2014)
34. P.C. Chang, J.L. Wu, J.J. Lin, A Takagi-Sugeno fuzzy model combined with a support vector regression for stock trading forecasting. Appl. Soft Comput. **38**, 831–842 (2015)
35. L.F. Chen, Z.T. Liu, M. Wu, M. Ding, F.Y. Dong, K. Hirota, Emotion-age-gender-nationality based intention understanding in human-robot interaction using two-layer fuzzy support vector regression. Int. J. Social Robot. **7**(5), 709–729 (2015)
36. Z. Liu, S. Xu, C.L.P. Chen, Y. Zhang, X. Chen, Y. Wang, A three-domain fuzzy support vector regression for image denoising and experimental studies. IEEE Trans. Cybern. **44**(4), 515–526 (2014)
37. X. Yang, G. Zhang, J. Lu, J. Ma, A kernel fuzzy c-means clustering-based fuzzy support vector machine algorithm for classification problems with outliers or noises. IEEE Trans. Fuzzy Syst. **19**(1), 105–115 (2011)
38. L. Bottou, V. Vapnik, Local learning algorithms. Neural Comput. **4**(6), 888–900 (1992)
39. K.P. Lin, A novel evolutionary kernel intuitionistic fuzzy c-means clustering algorithm. IEEE Trans. Fuzzy Syst. **22**(5), 1074–1087 (2014)
40. D.D. Nguyen, L.T. Ngo, Multiple kernel interval type-2 fuzzy c-means clustering, in *Proceedings of IEEE International Conference on Fuzzy Systems* (Hyderabad, India, 2013), pp. 1–8
41. W.M. Dong, F.S. Wong, Fuzzy weighted averages and implementation of the extension principle. Fuzzy Sets Syst. **21**(2), 183–199 (1987)
42. L.F. Chen, Z.T. Liu, F.Y. Dong, Y. Yamazaki, M. Wu, K. Hirota, Adapting multi-robot behavior to communication atmosphere in humans-robots interaction using fuzzy production rule based friend-Q learning. J. Adavanced Comput. Intell. Intell. Inform. **17**(2), 291–301 (2013)
43. L.F. Chen, Z.T. Liu, M. Wu, F.Y. Dong, Y. Yamazaki, K. Hirota, Multi-robot behavior adaptation to local and global communication atmosphere in humans-robots interaction. J. Multimodal User Interfaces **8**(3), 289–303 (2014)
44. S. Calinon, P. Kormushev, D.G. Caldwell, Pepper learns together with children: development of an educational application, in *Proceedings of IEEE-RAS 15th International Conference on Humaniod Robots* (Seoul, Korea, 2015), pp. 270–275
45. E. Bicho, W. Erlhagen, E. Sousa, et al., The power of prediction: robots that read intentions, in *Proceedings of IEEE/RSJ International Conference on Intelligent Robots and Systems* (Vilamoura, Portugal, 2012), pp. 5458–5459

Chapter 11
Emotion-Age-Gender-Nationality Based Intention Understanding Using Two-Layer Fuzzy Support Vector Regression

An intention understanding model based on two-layer fuzzy support vector regression (TLFSVR) is proposed in human-robot interaction, where Fuzzy C-Means clustering is used to classify the input data, and intention understanding is mainly obtained by emotion, with identification information such as age, gender, and nationality. It aims to realize the transparent communication by understanding customers' order intentions at a bar, in such a way that the social relationship between bar staffs and customers becomes smooth. To demonstrate the aptness of intention understanding model, experiments are designed in term of relationship between emotion-age-gender-nationality and order intention.

11.1 Introduction

As the gradual increase of the robotics community in social communication, robots that exist primarily to interact with people [1]. Humanoid robots, both cognitive and physical, become companions of humans, which open up an important research domain to develop human-robot interaction (HRI) [2]. Rapid advancement of robotic technology is bringing robots closer to everyday environments of people, such as home, school, and supermarket. Consequently, interaction between people and robots has become increasingly socially situated and multi-faceted [3]. Office roam-robot is designed to socially interact with people in daily lives [4] and socially assistive robot system is developed for elderly people in physical exercise [5], which indicate not only the benefits of having robots interact with people, but also the need for the interactions to be smooth and natural.

Robotic systems is developed to reflect the progress in different fields of robotics with regard to adaptation [6], learning [7], and autonomous behaviors [8]. Many

L. Chen et al., *Emotion Recognition and Understanding for Emotional Human-Robot Interaction Systems*, Studies in Computational Intelligence 926,
https://doi.org/10.1007/978-3-030-61577-2_11

researches conduct unveiling the importance of properly designed adaptive human-robot interaction strategies to adjust the robots' behaviors for smooth communication [9–11]. Still, HRI is often not so easy because robots lack understanding of the people's internal states, e.g., intentions, rather than just social behaviors, and vice versa.

Intention understanding is essential for many aspects of cognition, culture and social, including image intention understanding [12], speech intention understanding [13], and gesture intention understanding [14]. Although speech and gesture play important roles in intention understanding, they cannot intuitive reflect people's internal states. Emotion is an important missing link [15] in the intention-behavior gap, and it is essential for social behavior [16]. In addition, emotion is suggested as the driving force behind motivation on behavior. From a psychological perspective, emotion is more associated with people's internal state, i.e., intention. Therefore, emotion is supposed to be a key component to deeply reflect people's internal states, and how to use individual emotion for intention understanding becomes necessary and feasible, which is also a new challenge.

An intention understanding model is proposed to deeply comprehend people's inner thoughts according to the emotion, where age, gender, and nationality are additional informations for dealing with the problem of various people coexistence. Since unfamiliar issues becomes familiar through cognitive process, robot is firstly able to distinguish between familiar and unfamiliar people. In the case of familiar people, robot may understand his/her intention through memory, where the memory retrieval is based on the fuzzy production rules; in the case of unfamiliar people, an intention understanding model based on the two-layer fuzzy support vector regression (TLFSVR) is introduced to deeply comprehend people's inner thoughts. The TLFSVR is an extension of support vector regression (SVR) [17], in which two-layer consists of local learning layer and global learning layer. Local learning layer includes fuzzy c-means based input data classification and SVR based learning, and global learning layer outputs the intention of TLFSVR based on fuzzy weighted average algorithm. In view of the local learning [18], the training data is split into multiple subsets, so that the traditional one SVR is extended to multiple SVRs, for example, considering various people with different ages, different genders, and different nationalities, which may lead to the problem of health, gender hobby, religion, and so on. Different regression models reflect different ages/genders/nationalities, which is better than only one regression model covers all situations. Moreover, age, gender, and nationality play important role in intention understanding, for example they affect people's intention to purchase products or services [19], accept new things [20], and handle risky situation [21]. Thus, three components of human property, i.e., age, gender, and nationality, are used for intention understanding beside emotion, then training data can be classified according to the identification informations. Finally, fuzzy inference based on identification informations is used to generate the intention, which is either output of memory retrieval or the TLFSVR.

The proposal enables robot to deeply understand people's intention, and robot can transparently communicate with the user and guess what he/she expects. Transparent communication means one accesses the comprehension and memory for understanding the internal states of the other, e.g., intention [22, 23]. Transparently communicating to each other provides a new communication way which is from heart to heart, that is, our inner expectative intention is possible to carry out by emotion and some negative impact could be avoided based on different ages/genders/nationalities, which is a boost for smooth communication.

To show the effectiveness of the proposed intention understanding model, simulation and application experiments are done. Simulation experiment is designed to verify the accuracy of the proposal, where the simulation environment is built by Matlab in a PC employed with a dual core processer (2.8 GHz), memory (2.99 GB), and Windows 7 system. The training and testing data is collected from the information acquisition system, including the emotion recognition module and questionnaires module, which is created by C# in Visual Studio 2013. In addition, to confirm the validity and practicability of the proposed intention understanding model, application experiment of humans-robots interaction is conducted in the developing mascot robot system (MRS) [24], where a scenario of "drinking at bar" is performed by six salary men and two eye robots.

11.2 Two-Layer Fuzzy Support Vector Regression

SVR is being an extensive application in the field of human-robot interaction, and it meets some challenges, one is the difficulty to guarantee the good local modeling performance of the obtained model. Local learning algorithm is attempt to locally adjust the capacity of the model and based on the idea of local learning, two-layer fuzzy support vector regression (TLFSVR) is proposed to handle the evenly distributed problem that the training data sample cover the entire array.

11.2.1 Support Vector Regression

As a powerful and promising learning machine, support vector machines (SVMs) is famous for solving problems characterized by small samples, nonlinearity, high dimension, and local minima [25–27]. SVM employs structural risk minimization (SRM) principle to achieve better generalization ability, so that it provides higher performance than traditional empirical risk minimization (ERM) based learning machines, e.g., neural networks [28]. There are two main categories for SVM, which are support vector classification (SVC) and support vector regression (SVR). The goal of SVC is to construct an optimal separating hyperplane in a higher dimensional space by maximizing the margin between the separating hyperplane and classification data. Maximizing the margin is a quadratic programming (QP) problem and can be solved from its dual problem by introducing Lagrangian multipliers. Using the dot product functions called kernels, SVC finds the optimal hyperplane in feature space, which means kernel functions play a vital role in classification since they determine feature spaces where data sample are classified, and it may directly affect performances of the SVM classification. A combination of a few input points called support vectors can be used to generate the solution of the optimal hyperplane.

With the introduction of ε-insensitive loss function, the SVM is extended to solve regression problems, called support vector regression (SVR) [17]. While SVM is mostly used to perform classification by determining the maximum margin separation hyperplane between two classes, SVR tries the inverse, i.e., to find the optimal regression hyperplane so that most training samples lie within an ε-margin around this hyperplane. SVR is applied to various applications such as optimal control [29], time-series prediction [30], interval regression analysis [31]. Suppose training data set is a set of l pairs of vectors and the set is described as $D = \{(X_i, y_i) \,|X_i \in R^n, y_i \in R, i = 1, 2, \ldots, l\}$, Considering the simplest case of linear SVR, the basic function is given as

$$f(X) = \langle W, X\rangle + b \tag{11.1}$$

where the function $f(X)$ involves taking the dot product of the input vector X with a model parameter W and adding the scalar bias b.

The aim of ε-SVR is to find a function $f(X)$ that has at most ε deviation from the actually obtained targets y_i for all the training data, and at the same time as flat as possible by minimizing the norm of W. The problem can be described as convex optimization problem and written as

$$\begin{array}{ll} \min & \frac{1}{2}\,\|W\|^2 \\ \text{subject to} & y_i - \langle W, X_i\rangle - b \leqslant \varepsilon \\ & \langle W, X_i\rangle + b - y_i \leqslant \varepsilon. \end{array} \tag{11.2}$$

Sometimes constrains may not always be satisfied, that means not all pairs (X_i, y_i) fit with ε precision. Slack variables ξ_i and ξ_i^* are introduced to cope with the otherwise infeasible constrains, and (11.2) becomes

$$\begin{array}{ll} \min\limits_{W,b,\xi_i,\xi_i^*} & \frac{1}{2}\,\|W\|^2 + \gamma \sum\limits_{i=1}^{l} \left(\xi_i + \xi_i^*\right) \\ \text{subject to} & y_i - \langle W, X_i\rangle - b \leqslant \varepsilon + \xi_i \\ & \langle W, X_i\rangle + b - y_i \leqslant \varepsilon + \xi_i^* \\ & \xi_i, \xi_i^* \geqslant 0, i = 1, 2, \ldots, l \end{array} \tag{11.3}$$

where γ is a scalar regularization parameter that penalizes the amount of slack used (smaller values of γ yield more outlier rejection).

Retrieving the optimal hyperplane in (11.3) is a QP problem, which can be solved by constructing a Lagrangian and transformed into the dual as (11.4), where α_i and α_i^* are the Lagrange multipliers. All the dot product in (11.4) can be replaced by kernel function and the regression function is reconstructed as

$$\max_{\alpha,\alpha^*} \quad -\frac{1}{2}\sum_{i,j=1}^{l}\left(\alpha_i-\alpha_i^*\right)\left(\alpha_j-\alpha_j^*\right)\langle X_i, X_j\rangle - \varepsilon\sum_{i=1}^{l}\left(\alpha_i+\alpha_i^*\right)+\sum_{j=1}^{l} y_i\left(\alpha_i-\alpha_i^*\right)$$

$$\text{subject to} \quad \sum_{i=1}^{l}\left(\alpha_i-\alpha_i^*\right)=0 \ \ and \ \ \alpha_i,\alpha_i^*\in\left[0,\gamma\right] \tag{11.4}$$

$$f(X)=\sum_{i=1}^{l}\left(\alpha_i-\alpha_i^*\right)K(X_i,X)+b \tag{11.5}$$

where the kernel function is the Gaussian kernel,

$$K(x_i,x_j)=\exp\left(-\frac{(x_i-x_j)^2}{2\sigma^2}\right). \tag{11.6}$$

11.2.2 Two-Layer Fuzzy Support Vector Regression

SVR is being an extensive application [32] in the field of human-robot interaction, and it meets some challenges, one is the difficulty to guarantee the good local modeling performance of the obtained model, which means only one regression hyperplane cannot be guaranteed most training samples lie within it. For example, a training data sample is collected from different ages, considering that people in different ages may have their own hobbies, different regression models reflect different ages maybe better than only one regression model that involves all the ages. Local learning algorithm is attempt to locally adjust the capacity of the model to the properties of the training data subset [18], and based on the idea of local learning, two-layer fuzzy support vector regression (TLFSVR) is proposed to handle the evenly distributed problem that the training data sample cover the entire array. TLFSVR is an extension of the SVR, where local learning is fused with the SVR, as well as fuzzy logic, and two-layer includes local learning layer and global learning layer, as shown in Fig. 11.1.

First, the training data set are split into several subsets by using the fuzzy c-means (FCM) algorithm [33, 34]. The algorithm is based on the minimization of the following objective function,

$$f(U,V)=\sum_{i=1}^{l}\sum_{k=1}^{C}(\mu_{ik})^m\left\|X_i-V_k\right\|^2 \tag{11.7}$$

where $l\in N$ is the number of training data, $C\in[2,l)$ is the number of clusters, $m\in[1,+\infty)$ is a weighting exponent, $\mu_{ik}\in[0,1]$ is the degree of membership of training data X_i in the cluster k, V_k is the center of cluster k. Approximate optimization of (11.7) by the FCM algorithm is based on iteration, with the update of membership and the cluster centers V_k by

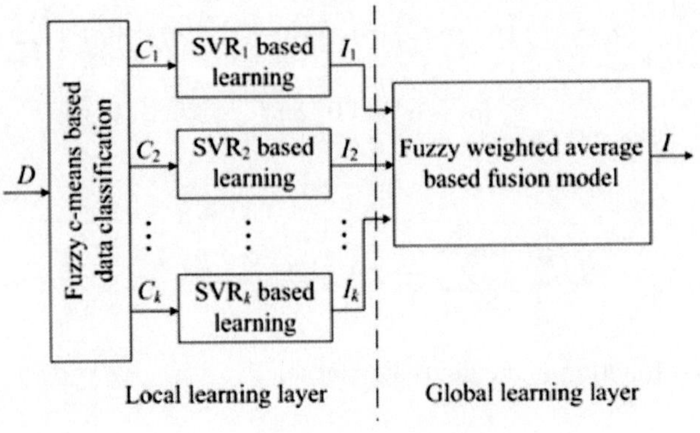

Fig. 11.1 Two-layer fuzzy support vector regression (TLFSVR)

$$\mu_{ik} = \left(\sum_{k=1}^{C} \left(\frac{d_{ij}}{d_{ik}} \right)^{\frac{2}{(m-1)}} \right)^{-1}, 1 \leqslant i \leqslant l, 1 \leqslant j \leqslant C \tag{11.8}$$

$$V_k = \frac{\sum_{i=1}^{l} (\mu_{ik})^m X_i}{\sum_{i=1}^{l} (\mu_{ik})^m}, 1 \leqslant j \leqslant C \tag{11.9}$$

where $d_{ik} = \|X_i - V_k\|$.

The iteration will stop when termination criterion is satisfied, i.e., $\|X_i - V_k\| \leqslant \varepsilon$, where $V = (v_1, v_2, \ldots, v_C)$, and ε is the given sensitivity threshold.

Then the spread width is calculated as

$$\delta_k^o = \sqrt{\frac{\sum_{i=1}^{l} (\mu_{ik})^m \left\| x_i^o - v_i^o \right\|^2}{\sum_{i=1}^{l} (\mu_{ik})^m}}, 1 \leqslant k \leqslant C, 1 \leqslant o \leqslant N. \tag{11.10}$$

According to the obtained centers and spread width, the training data sample is split into training subsets as (11.11), where η is a constant for controlling the overlap region of the training subsets and the size of training subsets increases as η increases, as well as computational time.

$$D_k = \left\{ (X_i, y_i) \left\| v_k^o - \eta \delta_k^o \leqslant x_i^o \leqslant v_k^o + \eta \delta_k^o \right. \right\}, 1 \leqslant i \leqslant l, 1 \leqslant o \leqslant N, 1 \leqslant k \leqslant C \tag{11.11}$$

Second, regression function of each cluster is constructed by (11.5) as follow

$$SVR_k = \sum_{i=1}^{l_k} \left(\alpha_{i,k}^* - \alpha_{i,k}\right) k\left(X_i, X\right) + b_k, X \in D_k, 1 \leqslant k \leqslant C \tag{11.12}$$

where l_k denotes the number of training data in the kth training subset, the parameters $\alpha_{i,k}^*$, $\alpha_{i,k}$ and b_k are obtained by the SVR for the kth training subset.

Finally, after the local learning, global learning for output is calculated based on the fuzzy weighted average algorithm [35] fusion model with triangle membership functions and SVRs. The membership function of the fuzzy weighted average algorithm is built as

$$\begin{aligned} A_k(X_i) &= a_k^1(x_i^1)a_k^2(x_i^2)\cdots a_k^o(x_i^o)a_k^o(x_i^o) \\ &= \max\left(\min\left(\frac{x_i^o(v_k^o-\eta\delta_i^o)}{v_k^o-(v_k^o-\eta\delta_k^o)}, \frac{(v_k^o+\eta\delta_k^o)-x_i^o}{(v_k^o+\eta\delta_k^o)-v_k^o}\right), 0\right), 1 \leqslant k \leqslant C \end{aligned} \tag{11.13}$$

where $A_k(X_i)$ is the fuzzy weight of the kth SVR.

Then, the global output of the proposed TLFSVR is based on the fusion model which is calculated using fuzzy weighted average algorithm as

$$y_i(X_i) = \frac{\sum_{k=1}^{C} A_k(X_i) SVR_k(X_i)}{\sum_{k=1}^{C} A_k(X_i)}, 1 \leqslant i \leqslant l, 1 \leqslant k \leqslant C. \tag{11.14}$$

11.3 Intention Understanding

Emotion based intention understanding is defined as estimating customers' order intentions according to customers' emotion at a bar, and emotion consists of facial expression and speech. Bar is a very popular place for both young people and salary man. At a bar, "what do you want to drink" is one of the customer's intention, if the bar staff knows very well about what is the right drink of customer in different emotion, it would be very nice service to the customer and they may come back again, especially for a salary man after work, a suitable drinks can release the heavy mood.

11.3.1 Emotion Based Intention Understanding

Intention can not be directly observable, but are inferred from expressive behavior, self-report, physiological indicators, and context. It is an important aspect of communication among people and is an essential component of the human cognitive system. The ability to understand the intent of others is critical for the successful communication and collaboration between people. In our daily interactions, we rely heavily on this skill, which allows us to "read" others' minds. If robots become effective in collaborating with humans, their cognitive skills must include mechanisms for inferring intention, so that they can understand and communicate naturally with people.

One way that a humanoid robot could facilitate intention understanding is based on emotion. Emotion helps inform and motivate social decision making, such as happiness, surprise, fear or anger inform us about the quality of our social relationships [36]. Emotion expression often promotes adaptive social responses in others. On the one hand, expression provides important coordinating information to other social partners [37]. Our reactions to events indirectly convey information about the evaluation of current desires and intentions, e.g., an undesired stimulus might result in an expression of disgust and an unexpected one might result in an expression of surprise. On the other hand, emotion expression seems to reflexively elicit adaptive social responses from others. Emotional behaviors are automatically alter perceptions and judgments through affective priming [38]. Emotion expression further triggers behavioral responses. For example, anger can elicit fear related responses or serve as a demand for someone to change their intention, sadness can elicit sympathy, and happy seems to invite social interaction.

Beside emotion, different drinks should be served according to customers' age and gender, in addition, nationality is also necessary to consider, for example, people are illegal to drink alcohol in Muslim country.

11.3.2 Characteristics Analysis

Using brainstorming, twenty-eight volunteers (i.e., different genders people aged 20–65 years, from 6 counties in Tokyo, as shown in Fig. 11.2) discuss topic like "what are the influence factors of intention understanding?", and the presentation of the brainstorming is shown in Fig. 11.3, where 9 questions are asked by volunteers during the brainstorming, questions/responses and the information of questioners/responders including age, gender, nationality, and religion are given.

According to the data analysis of brainstorming in Table 11.2, for influence factor "age", supporters are from different ages in 20–65; for influence factor "gender", supporters include both female and male; for influence factor "nationality", supporters are from 6 different countries; for influence factor "religion", supporters have different religions including Islam and Christian. Moreover, based on the basic information of volunteers in Table 11.4, which shows that they are different ages/genders/nationalities/educations/religions/occupations/income, and indicates the selected influence factors are signification. In addition, life experience, such as education, occupation, income, hobby, and wearing, are different with the increase of age, thus, age distribution is very important for selecting influence factors of intention, and a one-way ANOVA is used to see whether the supporters of each influence factor have age distribution. The null hypothesis is that all the supporters of each

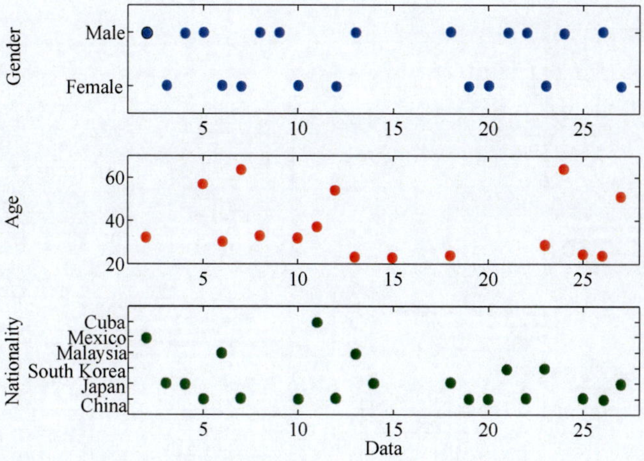

Fig. 11.2 Identification information of volunteers

influence factor have the same age distribution. Boxplots of age distribution by influence factors are shown in Fig. 11.4, according to analysis of variance in Table 11.1, the P-value is lower than 0.05, thus the null hypothesis can be rejected, that means supporters of each influence factor have age distribution. Since the signification of the 11 influence factors in Table 11.2, they are collected from the brainstorming.

As the discussion continues, the participants' passion slowly weaken, as well as the numbers. Moreover, the brainstorming is going to no-good situation, which the numbers of responders is less than half while discussing about the "weather", even under 40% in the discussion of "wearing". In this case, the brainstorming need to stop, and 9 factors which the support rate higher than 50% will be chose as initial influence factors of human intention understanding, including age, gender, nationality, education, emotion, religion, occupation, income, and hobby.

To select the independent factors for intention understanding model, first, twenty-eight volunteers (identification information are shown in Fig. 11.2) complete questionnaires about the correlation between collected relevant factors and the intention, and the correlation coefficient is from 0 to 1, e.g., correlation coefficient of age and order intention is 0.9, as shown in Table 11.3. The answer of each volunteer about the correlation between influence factors and intention are shown in Fig. 11.5. According to the basic information of volunteers in Table 11.4, it shows that they are different ages, different genders, different nationalities, different educations, different emotion (based on Pleasure-Arousal plane as shown in Fig. 11.9), different religions, different occupations, and different income, which indicate the responses are signification in reflecting the correlation between influences factors and intention.

Based on the correlation values in Fig. 11.5, nine intention related factors are classified into four clusters using the hierarchical clustering algorithm [39], as shown in Fig. 11.6.

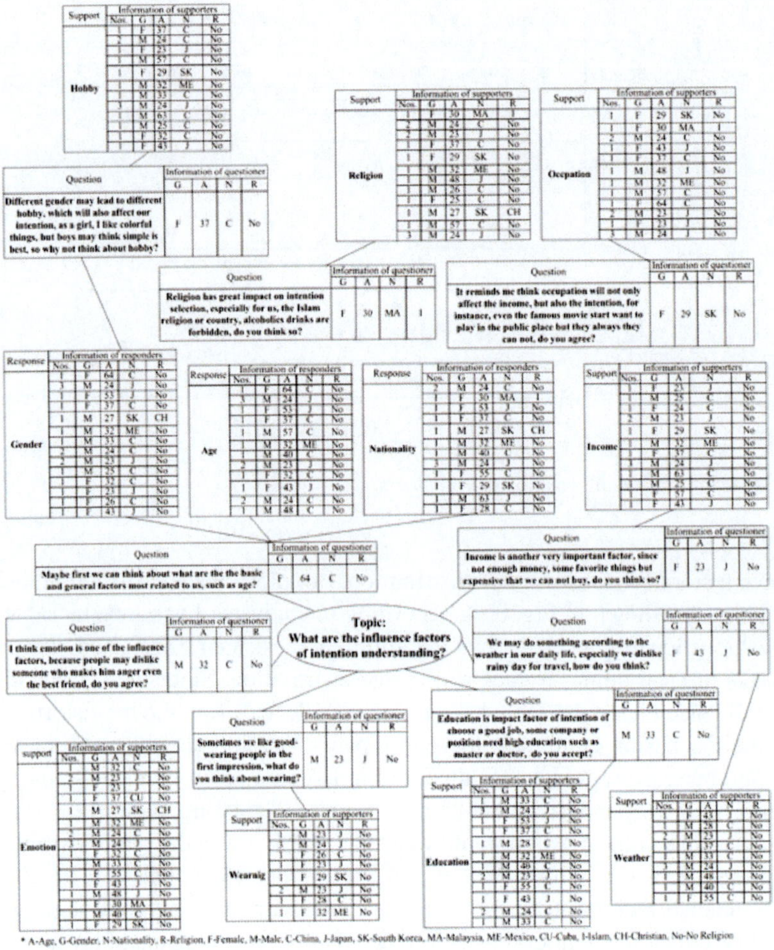

Fig. 11.3 Presentation of results of brainstorming

Figure 11.6 shows the dendrogram of factor classification, y axis is cluster distance from 0 to 1, relationship becomes week as the distance increase, with regard to this, 4 clusters are selected, where nationality and religion are placed in cluster 1; gender and hobby are classified in cluster 2; emotion is classified in cluster 3; age, education, occupation, and income are classified in cluster 4, and each cluster is relatively independent.

Table 11.5 shows the correlation coefficient between factors. Inputs of intention understanding model is selected based on the principle of minimum correlation and different clusters. According to Fig. 11.6 and Table 11.5, age, gender, nationality, and emotion are positioned in different clusters, and the correlation coefficient between each other are −0.0742 (age and gender), 0.0576 (age and nationality), −0.1761 (age and emotion), 0.0115 (gender and nationality), 0.1322 (gender and emotion), and 0.0517 (nationality and emotion), respectively, which indicates that age, gender, nationality, and emotion have little or no relationship to each other.

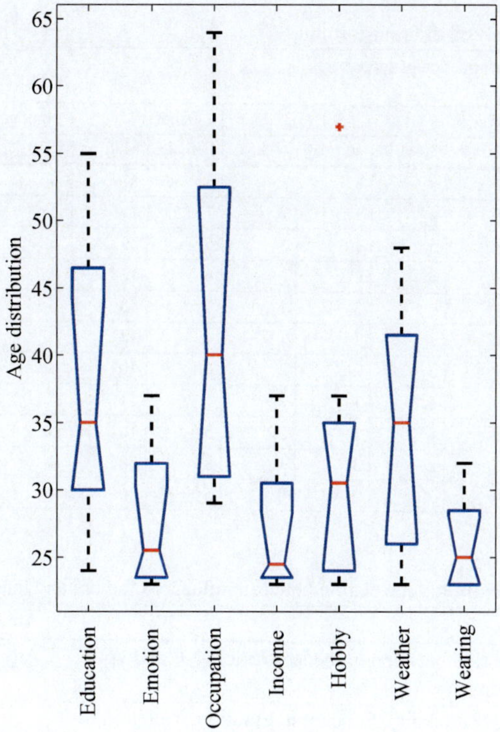

Fig. 11.4 Boxplots of age distribution by influence factors

Table 11.1 Analysis of variance

Source	SS	df	MS	F	$P_{rob} > F$
Columns	1801.96	6	300.327	3.17	0.004
Error	3961.75	49	80.852		
Total	5763.71	55			

Moreover, a one-way ANOVA is used to compare the means of different age/gender/nationality groups to determine whether they are significantly different from each other, and the observation parameter are correlation values between age/gender/nationality and intention in Fig. 11.5. According to the data sample in Table 11.6, 3 groups are classified based on different ages, 2 groups are classified based on different genders, 2 group are classified based on the nationality of China or Japan, and each group has 6 data. The null hypothesis is all the age/gender/nationality groups have equal influence on the intention. By using one-way ANOVA, boxplots of intention correlation by age/gender/nationality are shown in Fig. 11.7, respectively. According to ANOVA summary in Table 11.7, all the P-values of 3 samples are lower than 0.05, thus the null hypothesis is rejected, that means age, gender, and nationality have signification influence on the intention. Since the independence and

Table 11.2 Result analysis of brainstorming

Influence factor	Support rate (%)	Information of supporters								
		Age				Gender		Nationality	Religion	
		<30	30 ~ 40	40 ~ 50	> 50	Female	Male		Islam	Christian
Age	57.1	7	3	3	3	5	11	4	0	0
Gender	64.3	11	4	1	2	7	11	5	0	1
Nationality	53.6	8	3	1	3	6	9	6	1	1
Education	57.1	8	4	2	2	4	12	4	0	0
Emotion	71.4	10	6	3	1	7	13	6	1	1
Religion	57.1	11	3	1	1	4	12	6	1	1
Occupation	57.1	9	3	2	2	6	10	6	1	0
Income	53.6	10	2	1	2	5	11	5	0	0
Hobby	53.6	8	4	1	2	5	10	5	0	0
Weather	42.9	6	2	3	1	3	8	3	0	0
Wearing	39.3	10	1	0	0	5	6	4	0	0

Table 11.3 Questionnaire for correlation between influences factors and intention

Questions	Answers
How big is the correlation between influences factors IF and your intention to do something?	Any number from 0 to 1
($IF \in$ {Age, Gender, Nationality, Education, Emotion, Reliction, Occupation, Income, Hobby}	

Table 11.4 Basic information of volunteers

Influence factor	Cluster	Numbers	Influence factor	Cluster	Numbers
Age	<30	14	Education	Master	13
	30 ~ 40	6		Doctor	15
	40 ~ 50	3	Religion	Islam	1
	>50	5		Christian	1
Gender	Female	11		No	26
	Male	17	Occupation	Student	22
Nationality	China	13		Teacher	3
	Japan	10		Salary men	3
	South Korea	2	Income	Scholarship	12
	Malaysia	1		Salary	6
	Mexico	1		No	10
	Cuba	1			

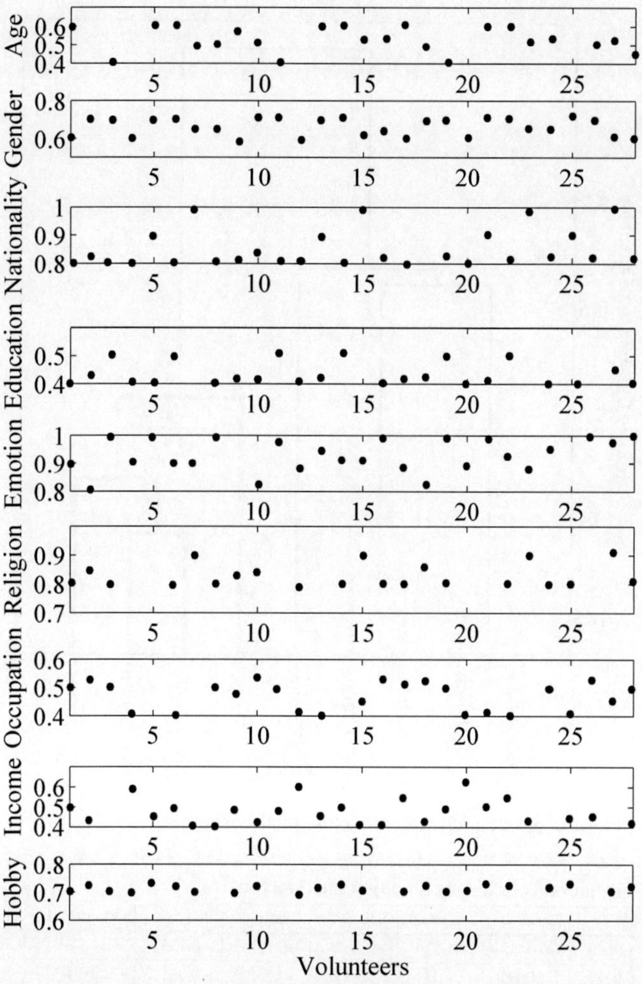

Fig. 11.5 Correlation between influence factors and intention

signification of age, gender, nationality, and emotion, they can be chose as the inputs of the intention understanding model.

In addition, age is selected since not only it is the accumulation of changes in a person over time, where education, occupation, and income change with the age, but also some legal age existed, such as drinking age and driving age, voting age and so on; gender refers to female and male, it is selected because a broader set of hobby can be obtained from the gender point of view; nationality is selected rather than religion, because it indicates the legal relationship between a person and a state, and the person should obey the state law, such as people are illegal to drink alcohol in some Muslim country, in such a way that, nationality includes religion. With these in mind, age, gender, nationality and emotion are suitable inputs for the intention understanding model.

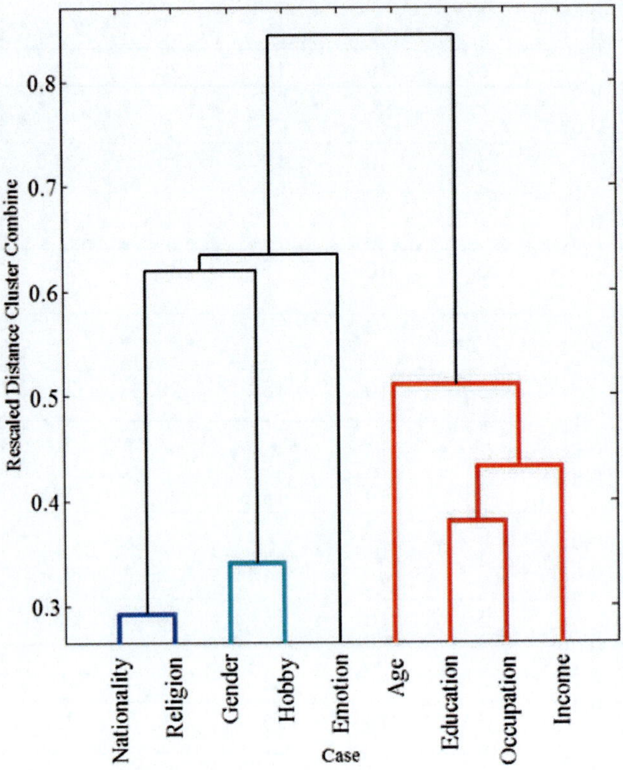

Fig. 11.6 Dendrogram of intention related factors relationship

Table 11.5 Correlation coefficients among related factors

	Age	Gender	Nationality	Education	Emotion	Religion	Occupation	Income	Hobby
Age	1.0000	−0.0742	0.0576	−0.2075	−0.1761	−0.1556	−0.4176	0.1355	0.2825
Gender	−0.0742	1.0000	0.0115	0.4301	0.1322	−0.0871	−0.0191	−0.2819	0.6820
Nationality	0.0576	0.0115	1.0000	0.0082	0.0517	0.7508	−0.2641	−0.4680	0.1320
Education	−0.2075	0.4301	0.0082	1.0000	0.0527	0.0986	−0.0999	0.0834	0.0351
Emotion	−0.1761	0.1322	0.0517	0.0527	1.0000	−0.2952	−0.0737	−0.1493	−0.0404
Religion	−0.1556	−0.0871	0.7508	0.0986	−0.2952	1.0000	0.1847	−0.5078	0.0043
Occupation	−0.4176	−0.0191	−0.2641	−0.0999	−0.0737	0.1847	1.0000	−0.4705	−0.1578
Income	0.1355	−0.2819	−0.4680	0.0834	−0.1493	−0.5078	−0.4705	1.0000	−0.3308
Hobby	0.2825	0.6820	0.1320	0.0351	−0.0404	0.0043	−0.1578	−0.3308	1.0000

Table 11.6 Data samples of ANOVA

Independent factor	Correlation between factor and intention						Information of responders					
Age	0.45	0.4	0.48	0.4	0.5	0.4	24	25	23	26	24	29
	0.5	0.51	0.49	0.52	0.53	0.51	32	33	32	37	32	30
	0.6	0.61	0.6	0.58	0.6	0.61	63	57	48	40	55	64
Gender	0.58	0.6	0.62	0.59	0.64	0.62	Male					
	0.71	0.7	0.72	0.65	0.71	0.65	Female					
Nationality	0.81	0.82	0.8	0.82	0.825	0.8	China					
	0.9	1	0.9	0.82	0.9	0.835	Japan					

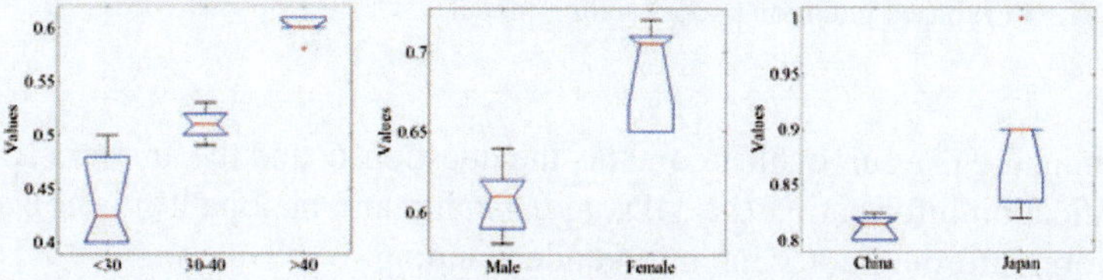

(a) Boxplots of correlation by age (b) Boxplots of correlation by gender (c) Boxplots of correlation by nationality

Fig. 11.7 Boxplots of intention correlation by age/gender/nationality

Table 11.7 ANOVA summary of (a)/(b)/(c)

Source	SS	df	MS	F
Columns	0.07874/0.02001/0.0192	2/1/1	0.03937/0.02001/0.0192	50.55/26.74/9.2
Error	0.01168/0.00748/0.02088	15/10/10	0.00078/0.00075/0.00209	
Total	0.09043/0.02749/0.04008	17/11/11		

11.4 Intention Understanding Model

Based on the TLFSVR and characteristics analysis, an intention understanding model is proposed to comprehend peoples' inner thoughts in human-robot interaction as shown in Fig. 11.8. The understanding model reflects empirical findings accumulated in cognitive (memory), at the same time, machine learning and fuzzy logic indicate the relationship of human intention and emotion. The basic idea is that the intention understanding model may support the understanding based on the emotion, where age, gender, and nationality are the additional informations. It mainly includes four modules, i.e., ID mapping, fuzzy rules based memory retrieval, two-layer fuzzy support vector regression (TLFSVR) based intention understanding, and fuzzy inference based intention generation. In addition, the information acquisition consists of emotion recognition and questionnaire. Emotion recognition is a supported module for intention understanding, which can read people's emotion from facial expression and speech.

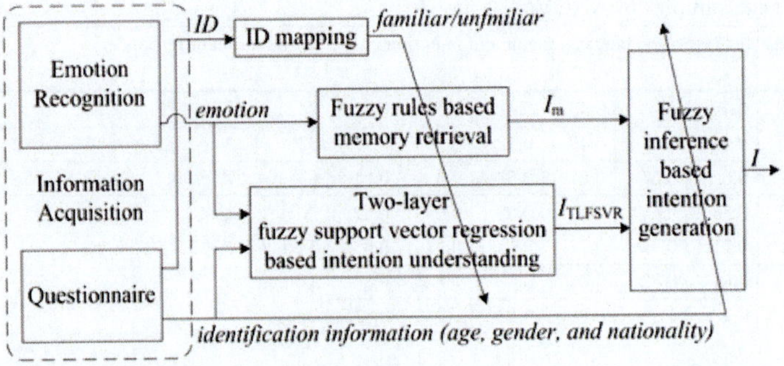

Fig. 11.8 TLFSVR based intention understanding model

Questionnaire is a survey that contains the question Q and the answers A, where the identification information (i.e., ID, age, gender, and nationality) is collected as identification information for intention understanding.

ID mapping module is used to distinguish between familiar and unfamiliar people based on the identity card (ID). Since unfamiliar issue usually becomes familiar through cognition process in human life, to some extent, people can be classified into familiar and unfamiliar.

Fuzzy rules based memory retrieval module aims to understand familiar people's intention directly from memory.

TLFSVR based intention understanding module is a SVR learning machine fusion with fuzzy logic, which determines the mapping from emotion to intention with the identification information, i.e., age, gender, and nationality.

Fuzzy inference based intention generation module is based on the identification informations (i.e., age, gender, and nationality), which is fuzzy selection operation from the output of memory retrieval module and TLFSVR based intention understanding module by using "IF-THEN" fuzzy rules.

11.4.1 Emotion Recognition

For understanding people's intention, emotion recognition is a supportive module based on the previous study [40, 44], which is realized using weighted fusion fuzzy inference based on the bimodal information, i.e., facial expression and speech. Emotion is recognized by fusing semantic cues from speech and facial expression, and represented in the Pleasure-Arousal plan, where happiness, surprise, fear, anger, disgust, sadness, and neutral are employed as the basic human emotion, with a total of 25 kinds of emotions, as shown in Fig. 11.9.

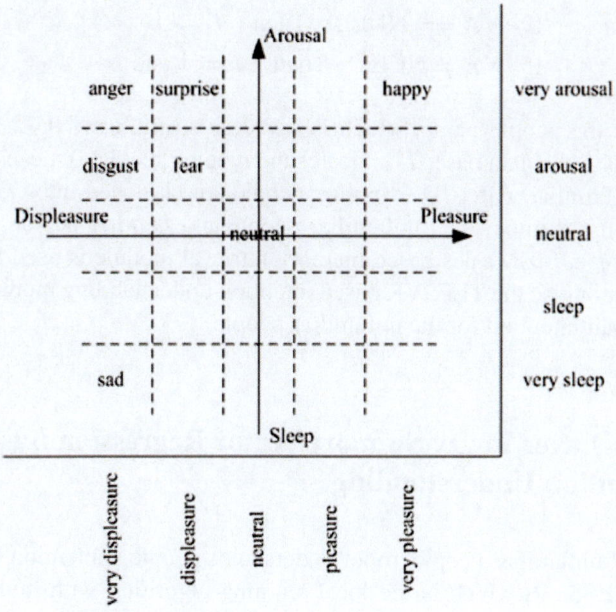

Fig. 11.9 Pleasure-Arousal emotion plan

11.4.2 Questionnaire

To collect the identification information (i.e., ID, age, gender, and nationality) for intention understanding, questionnaire is designed, which consists of a number of questions *Q* and their corresponding answers *A*.

In the questionnaire, questions *Q* contains "what is your name? / how old are you? / what is your gender? / what is your nationality?", and answers *A* are close ended (responding by a specific piece of information), for example in the simulation, volunteers need to write down the name/age/gender/nationality in the information acquisition module in Fig. 11.11 and then answers are collected as shown in Fig. 11.13. More specifically, ID is used to distinguish the people between familiar and unfamiliar in the human-robot interaction, while age, gender, and nationality are the identification informations for intention understanding.

11.4.3 ID Mapping

ID mapping is designed to transform the people name into different types of people, and realized by fuzzy rules such as

$$\begin{aligned}&\text{If}\quad NA == \text{NULL}\quad \text{Then}\quad T = 0\\&\text{If}\quad NA \neq \text{NULL}\quad \text{Then}\quad T = 1 \text{ and } NA\end{aligned} \tag{11.15}$$

where NA is the people name, which is saved as string, and NULL means NA (not involved) in the data base; T indicates the type of people, 0 means unfamiliar, and 1 means familiar. After ID mapping, people are classified into two categories according to the memory, i.e., unfamiliar people and familiar person. If he/she is a familiar people, fuzzy rules based memory retrieval module is used for intention understanding or else the TLFSVR based intention understanding module is used to generate human intention for the unfamiliar people.

11.5 Two-Layer Fuzzy Support Vector Regression Based Intention Understanding

In the case of unfamiliar people, robot understands people's intention based on the introduced TLFSVR, which is the local learning algorithm with multiple regression functions. Using the TLFSVR, robot can handle the multiple classes of input, where different regression functions corresponding to different classes of input. The TLFSVR based learning adaptation mainly consists of 7 steps:

Step 1: Initialization of FCM clustering

Parameters, i.e., the number of clusters C, the sensitivity threshold ε, and the overlap constant η are set.

Step 2: FCM calculation

The membership u_{ik} is calculated by (11.8), the cluster centers V_k is calculated by (11.9), and the spread width δ_k^o is calculated by (11.10).

Step 3: If the stop criterion has not satisfied, then go to *Step* 2; otherwise, go to *Step* 4.

Step 4: FCM clustering

The training subsets D_k are obtained by (11.11) based on the cluster centers and spread width.

Step 5: Initialization of SVR

Parameters, i.e., the Gaussian kernel width σ and the scalar regularization parameter γ are set.

Step 6: SVR learning

Regression function of each cluster SVR_k is constructed by SVR learning approach.

Step 7: Intention output

Intention understanding of the TLFSVR is outputted based on the fuzzy weighted average algorithm by using (11.14).

Table 11.8 Look-up table for output

Age	Nationality	
	Islam country or religion	Others
$<A_{lower}$	Non-alcoholic drinks	Non-alcoholic drinks
$A_{lower} \sim A_{upper}$	Non-alcoholic drinks	I_m/I_{SVR}
$>A_{upper}$	Non-alcoholic drinks	Non-alcoholic drinks

11.5.1 Intention Generation by Fuzzy Inference

Based on the identification informations, i.e., age, gender, and nationality, output of intention understanding model is designed by the look-up table, as shown in Table 11.8.

Where A_{lower} is the legal drinking age according to different countries, e.g., 18 in China, 20 in Japan, 19 in South Korea, and so on. A_{upper} is the healthy drinking age, e.g. low-risk drinking style for healthy life is begin from age 65 [41].

11.6 Memory Retrieval for Intention Understanding

Since unfamiliar issue usually becomes familiar in human cognition process, people may be classified into two cases, i.e., unfamiliar people and familiar people. Memory retrieval module is used to generate intention in the case of familiar people. The memory module stores the personal information of people, e.g., name, emotion, and intention. The fuzzy production rules of the memory rule base are Mamdani fuzzy rules designed as

$$R_{i\times j\times k}: \quad \text{If} \quad x_1 == NA_i \text{ and } x_1 == E_{ij} \quad \text{Then} \quad I_{jk} \text{ with } GR_{jk} \tag{11.16}$$

where $i \times j \times k$ is the number of rules, x_1 and x_2 are inputs, NA_i is the ith people name, E_{ij} is jth emotion of ith people, I_{jk} is kth intention in jth emotion of ith people, and GR_{jk} is the generation rate of the corresponding intention. For example, a scenario of "drinking at a bar" in human-robot interaction, suppose robot is a bartender and a familiar customer named FC with happy emotion, who drinks the beer when felt happy last time, before ordering, since memory retrieval function of the robot, one of the fuzzy rules has been selected, which is defined as "If $x_1 =$ FC and $x_2 =$ happy Then $I =$ beer with $GR = 1$".

Several intentions may be suitable for the expected emotion, taking into account this problem, fuzzy rule is selected based on the generation rate of the intention. If people produce a new intention to the expected emotion, then a rule is added to the memory rule base. Subsequently, the generation rate ($GR \in [0, 1]$) of the rule is updated when the same rule is used again and is calculated in terms of time-weighted statistics as

$$GR = \frac{UT}{UT + OT} \tag{11.17}$$

where $UT \in N$ is the number of use times of kth intention in jth emotion, and $OT \in N$ is the number of use times of other intentions in jth emotion. Adding or deleting fuzzy rules depend on two thresholds, i.e., the upper bound ($B_{upper} = 1$) and the lower bound ($B_{lower} = 0.8$). The generation rate of effective rule must be between these two boundaries, otherwise, it will be deleted.

11.7 Experiments

In this part, the data settings and experimental environment settings are explained. The simulation experiment is carried out and the experimental results are analyzed. To validate the proposed intention understanding model, simulation experiment and application experiment are both done. Both simulation and application choose the scenario of "drinking at a bar" as the background, this is due to the prevalence of the bar culture in the word, especially for youth and salary man.

11.7.1 Experiment Setting

In Japan, bar is endowed with a new form that is a type of Japanese drinking establishment, called Izakaya. Izakaya is a casual places of after-work drinking and very popular for the salary man, even becomes a part of their life. If the owner of an Izakaya understands each customer preferences, e.g., a right drink for specific mood, it would be very nice and deeply touched the tired salary man, by which it will also increase the turnover. The customer preferences problem seems to be a very interesting research, so the goal of the simulation and application experiments is understanding of the customer's intention to order at a bar based on the customer's emotion.

Taking the complexity of intention understanding process into account, simulation experiments are first performed, which not only can run repeatedly to simplify debugging of the intention understanding process, but also can simulate various emotions that are seldom experienced in real life. Simulation experiment is designed to verify the accuracy of the proposal, and is conducted in the Matlab environment, which is ran in a PC with dual core processer (2.8 GHz), memory (2.99 GB), and Windows 7 system. The training and testing data is collected from the information acquisition system, including the emotion recognition module and questionnaires module, which is created by C# in Visual Studio 2013.

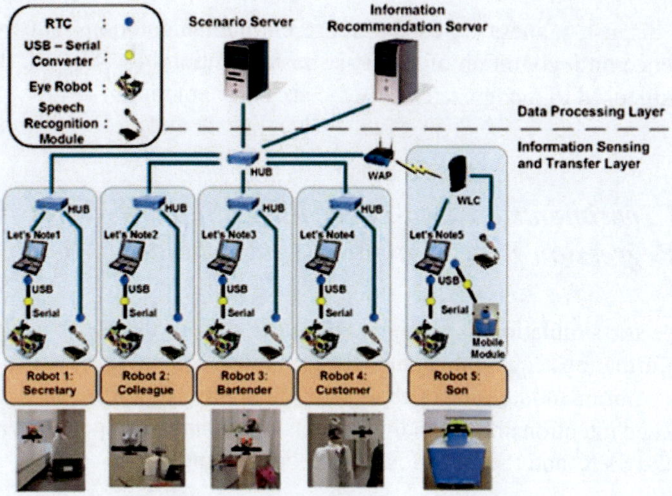

Fig. 11.10 Network system of MRS

The application experiment is developed to confirm the validity and practicability of the proposed intention understanding model in the developing Mascot Robot System (MRS) [24]. MRS is a typical application of humans-robots interaction, which is an information presentation system destined to be used mainly as an information terminal for communication with humans. MRS is now developing to the salary man's life for communication between robot and salary man [42].

In the MRS, five eye robots play different roles in the life of salary man, i.e., four fixed eye robots act as secretary, colleague, bar lady, and customer, respectively; an autonomous mobile eye robot acts as son of the salary man. Robot technology middleware (RTM) is used to construct the network system of MRS called RTM-Network, where each robot can be viewed as a network component and each function unit of every robot is called robot technology component (RTC), as shown in Fig. 11.10.

RTM-Network is a two-layer network system, including the information sensing and transfer layer as well as data processing layer. There are eight modules in the RTM-Network, including the speech recognition module (SRM), facial expression recognition module (FERM), gesture recognition module (GRM), eye-robot control module (ERCM), emotion processing module (EPM), display module, mobile robot control module, scenario server module, and information retrieval module [43]. The SRM sensing the speeches information through microphone, the FERM and GRM sensing the facial expressions and gestures information using Kinect. Moreover, the EPM obtains the recognized results from the SRM, the FERM, and the GRM, and then transmits them to the server module for the further processing, such as the calculation of individual emotions. Finally, each robot understands people's intention according to the obtained emotion. In the preliminary application, the FERM is mainly used for emotion recognition, and only six basic emotion including happiness, surprise, fear, anger, disgust, sadness are used to represent human emotions, since MRS has rich experience on recognition of these six basic emotion [40, 42, 43], the FERM module is supposed to recognize exactly the six basic emotion.

11.7.2 *Experiments on Two-Layer Fuzzy Support Vector Regression Based Intention Understanding Model*

Data sample for simulation experiment is collected from thirty-two volunteers by using the information acquisition system. Information acquisition system includes the emotion recognition module and questionnaires module. To validate the performance of the proposed intention understanding model, a comparative experiment among the TLFSVR, the SVR, and the BPNN is carried out in simulation.

11.7.2.1 Data Preparation

Data preparation is designed based on the scenario of "drinking at a bar". Data sample for simulation experiment is collected from thirty-two volunteers by using the information acquisition system, as shown in Fig. 11.11. Information acquisition system is created by C# in Visual Studio 2013, which includes the emotion recognition module and questionnaires module.

Questionnaires are about the relationship between emotion and intention to order at a bar, at the same time, write down the identification information, such as age, gender, nationality, as shown in Table 11.9. For easy understanding and selecting, questionnaire is transformed into questionnaires module in Fig. 11.11, including the identification information and order intention selection. The order intentions mainly include the top six popular drink in the Izakaya.

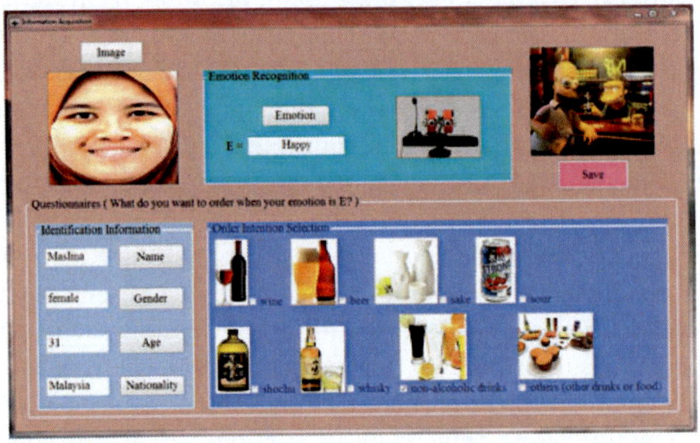

Fig. 11.11 Information acquisition

Table 11.9 Questionnaire for relationship between emotion and intention to order at a bar

Questions	Answers
What do you want to order when your emotion is E? ($E \in$ {anger, disgust, sadness, surprise, fear, happiness, etc.}, 25 questions corresponding to 25 kinds of emotions in total, as shown in Fig. 11.9)	1-Wine, 2-beer, 3-sake, 4-sour, 5-shochu, 6-whisky, 7-non-alcoholic drink, 8-others (other drinks or food)

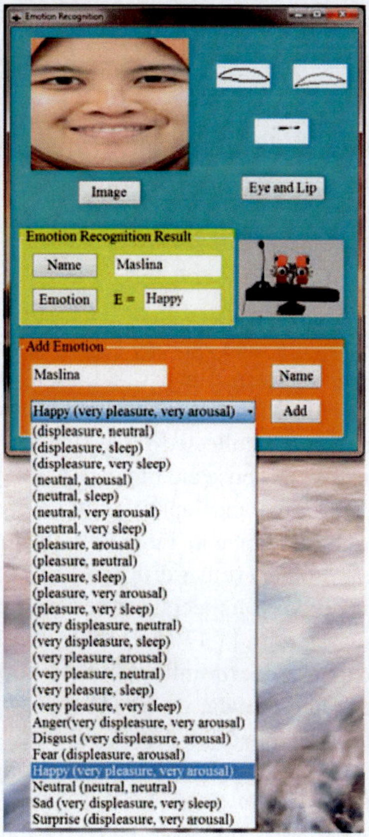

Fig. 11.12 Emotion recognition

Emotion recognition is realized by facial expression, where twenty-five kinds of emotions in total according to the Pleasure-Arousal emotion space, as shown in Fig. 11.12. By using the emotion recognition module, it can not only directly give the recognition result, but also can add any one of the 25 kinds of emotion and the corresponding facial expression states to the Access 2013 database. Since the emotion recognition module has the function of emotion recording, in the case of having the facial expression data records of volunteers, emotion recognition accuracy is guaranteed and will not affect the intention understanding accuracy.

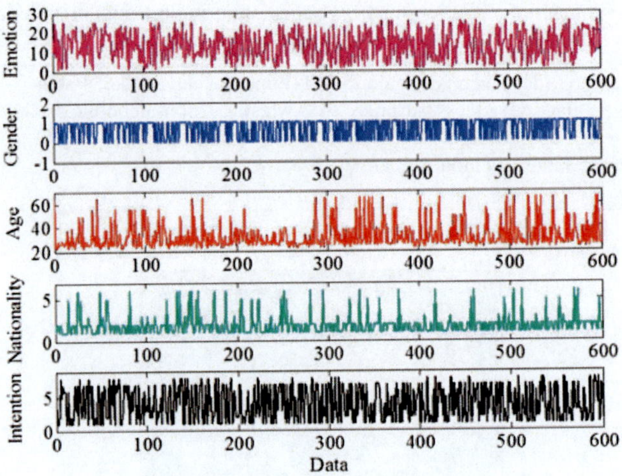

Fig. 11.13 Training data for simulation

Volunteers contains five salary men aged 30 to 65 years old and seventeen postgraduate students aged 20 to 35 years. They are different genders, and six countries (i.e., China, Japan, South Korea, Malaysia, Cuba, and Mexico) in Tokyo. Moreover, to get valid results, more data are collected by asking ten actual customers at the Izakaya named Watami in Tokyo. Those are of aged 20–60 years, including 6 males and 4 females, mainly from China and Japan. They ordered randomly, and if it is beyond the popular six drinks as shown in Table 11.9, the intention will be collected as 7-non-alcoholic drink or 8-others (other drinks or food). A total of 800 groups of data are collected from thirty-two volunteers, where 600 groups for training and 200 groups for testing, as shown in Figs. 11.13 and 11.14.

N-way ANOVA is used to examine the influence of emotion/age/gender/nationality on intention by using the collected data, where the independent variables are emotion, age, gender, and nationality, and the dependent variable is intention. The null hypothesis is all the emotion/age/gender/nationality groups have equal influence on the intention. According to analysis of variance reported in Table 11.10, all the P-values of 4 independent variables are lower than 0.05, thus the null hypothesis can be rejected, that means emotion, age, gender, and nationality have signification influence on the intention. Since the collected data can display the statistical significance of emotion, age, gender, nationality, they can be applied to the simulation study.

11.7.2.2 Experimental Results

To validate the performance of the proposed intention understanding model, a comparative experiment among the TLFSVR, the SVR, and the BPNN is carried out in simulation, aiming to generate the intention by emotion and discuss the influence of age/gender/nationality on the intention understanding.

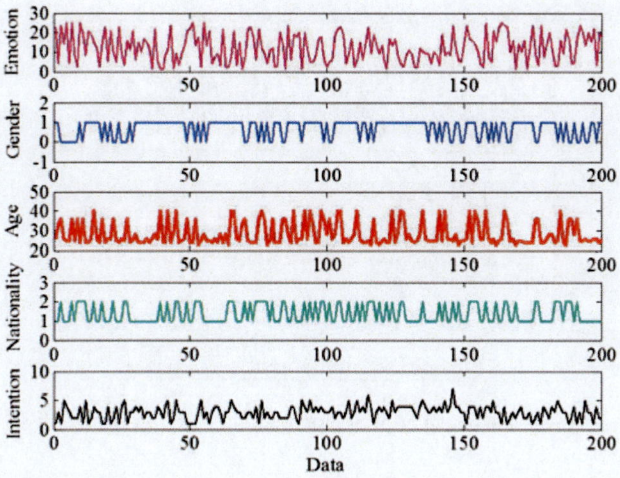

Fig. 11.14 Testing data for simulation

Table 11.10 Analysis of variance

Source	Sum Sq.	d.f.	Mean Sq.	F	$P_{rob} > F$
Emotion	492.62	24	20.256	5.92	1.1102×10^{-16}
Age	763.79	13	58.753	16.93	0
Gender	111.44	1	111.435	32.12	2.0632×10^{-8}
Nationality	484.08	3	161.359	46.51	0
Error	2626.5	757	3.47		
Total	4186.87	799			

Simulation experiment concerns a situation of "drinking at the bar" based on the collected data of emotion and order (intention), as well as the identification information (i.e., age, gender, and nationality). The goal is to realize the transparent communication (communication from heart to heart), that is, our inner expectative intention is possible to carry out, and is a boost for smooth communication. In other words, correctly obtaining the customer' order by using the proposed intention understanding model base on the customer's emotion and identification information. According to the goal, four indexes are used to evaluate the performance, including accuracy, computational time, average absolute relative deviation (AARD), correlation coefficient, where AARD is defined as

$$\%AARD = \frac{100}{N} \sum_{i=1}^{N} \left| \frac{y_{cal} - y_{exp}}{y_{exp}} \right| \tag{11.18}$$

where N denotes the number of the testing data, y_{cal} represents the output of the proposal, y_{exp} is the actual value from information acquisition system.

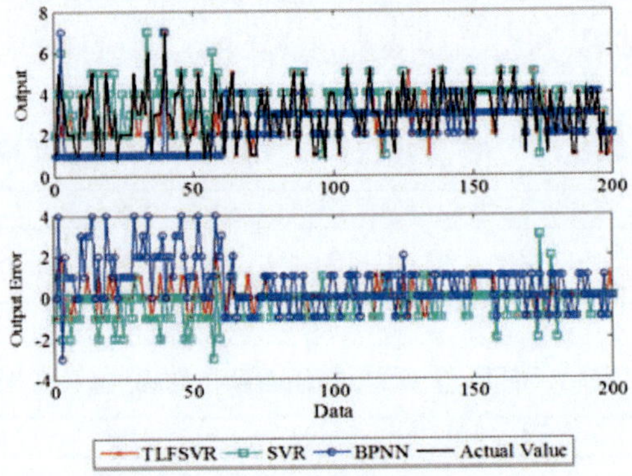

Fig. 11.15 Output curves and output error of proposed TLFSVR, SVR, and BPNN with $C = 2$

In the first experiment, the number of clusters C for the FCM is chosen as 2 according to the gender (i.e., female and male). In the second experiment, the number of clusters C for the FCM is chosen as 3 according to the age distribution (in general, people aged <30 belong to student, aged 30–55 belong to worker, and age >55 belong to retire). In the third experiment, the number of clusters C for the FCM is chosen as 6 according to the nationality (volunteers for data collection are from 6 different countries). Generally speaking, the overlap parameter η can be selected as 2.5, 3.5, 4, and 5 corresponding to the dimensionality of input variables are 1, 2, 4, and 8, respectively. According to the input variables are 4 in the experiment, the overlap constant η and the sensitivity threshold ε for the FCM are set to 3 and 1, respectively. For each SVR, the Gaussian kernel width σ is suggested as (0.1 0.5) $\times$ (rang of the inputs), since the inputs are all normalized, σ is chosen as 0.5, and the scalar regularization parameter γ are chosen as 200. BPNN has three layers (input layer, hidden layer, and output layer units are 4, 12, and 1, respectively), where learning rate is 0.25, and inertia coefficient is 0.05.

Figures 11.15, 11.16, and 11.17 show the output curves and output error of the proposed TLFSVR, SVR, and BPNN with $C = 2, 3, 6$, respectively, and the data analysis of result comparison among the three methods is showed in Table 11.11. According to simulation results, while C = 2/3/6, the proposed TLFSVR obtains an accuracy of 70%/72%/80%, which is 23%/26%/33% and 35.5%/37.5%/45.5% higher than SVR and back propagation neural networks (BPNN), respectively. Computational time is about 0.976s/0.935s/0.67s by using the proposed intention understanding model, indicating that the computing speed is about 1.9/2/2.8 and 3.6/3.7/5.2 times faster than the SVR and BPNN, respectively.

In the simulation examples, it can be found that the accuracy of the proposed understanding model is higher than that of the SVR and the BPNN, leading to the smaller AARD and higher correlation coefficient between the output and the actual value. Smaller AARD and higher correlation indicate that the TLFSVR based intention understanding model better matches the actual situation. The increase of the number of clusters number (C) improves the understanding accuracy to some extent.

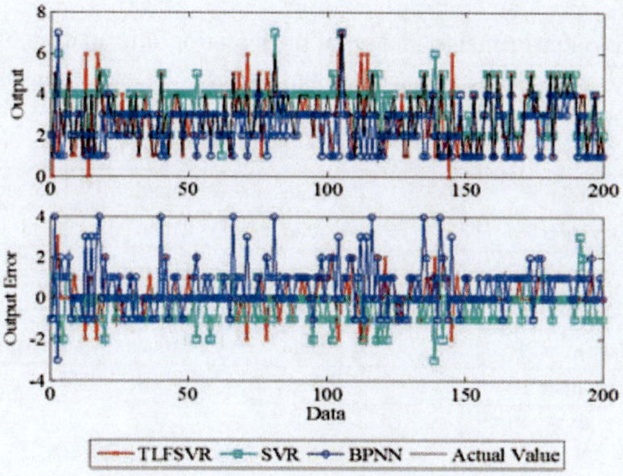

Fig. 11.16 Output curves and output error of proposed TLFSVR, SVR, and BPNN with $C = 3$

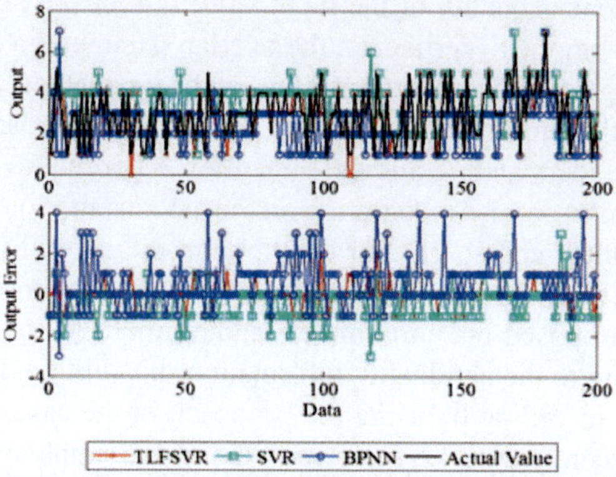

Fig. 11.17 Output curves and output error of proposed TLFSVR, SVR, and BPNN with $C = 6$

Table 11.11 Result comparison of simulation for intention understanding

Method		Index			
		Accuracy (%)	Computational time (s)	AARD (%)	Correlation
TLFSVR	C = 2	70	0.976	15.450	0.869
	C = 3	72	0.935	10.417	0.915
	C = 6	80	0.67	10.405	0.928
SVR		47	1.889	31.208	0.771
BPNN		34.5	3.505	33.767	0.456

Table 11.12 Identification information and emotion-intention information of volunteers

Group	ID	Identification information			Emotion	Intention of drink
		Age	Gender	Nationality		
Age-different	Salary man A	24	Male	China	Anger	Beer
	Salary man B	33	Male	China	Sad	Sake
Gender-different	Salary man A	25	Female	Japan	Happy	Whisky
	Salary man B	25	Male	Japan	Disgust	Sour
Nationality-different	Salary man A	28	Male	China	Surprise	Sour
	Salary man B	28	Male	Japan	Fear	Shochu

The overlap parameter ε is an important parameter in the proposed understanding model, which becomes larger as the dimensionality of the training data becomes higher. Since the dimensionality of the input value is four (i.e., emotion, age, gender, and nationality) in the experiments, the overlap parameter ε can be set large to cover all the training data, so that the accuracy can be guaranteed. Moreover, besides emotion, identification information including age, gender, and nationality are added as supplement for understanding humans' intention according to Table 11.12, especially to deal with miss or false recognition on the hidden emotion such as "curious", "forced smile", "cute", and so on. For example, in the case of people from Islam country, whatever the emotion, miss or false, they will not choose alcoholic drinks, and by using the proposed intention understanding model, a non-alcoholic drinks is selected according to the nationality information. In addition, the ID information (people's name) is recording by using the proposal, in the case of familiar ID, the favorite drink (frequently order) is selected. All of these supplementary information are used to reduce the impact of miss or false emotion recognition and guarantee the intention understanding accuracy to some extent. The computational time mainly includes some processes such as the split of the training data into C training subsets using the FCM clustering algorithm, constructing of all SVRS, and the calculation of the output with a fuzzy weighted average algorithm. The computational time is affected by the computational complexity of the proposal. The computational complexity is explicitly expressed in terms of the number of data and computation scale, for example, n is the number of the training data, n_i is the number of the training data in the ith training subset. C is the number of clusters, and m is the numbers of layers in BPNN, then computational complexity of the SVR is set as $O(n^2)$; for the proposal, there are C times of the SVR conducted with d_i training data, that is $O(\sum_{i=1}^{C} n_i^2)$; the computational complexity is $O(m \times n^2)$ for BPNN. Since computational complexity of the proposal is the simplest among these three methods, the computational time of it is less than that of the SVR and BPNN.

11.8 Summary

A two-layer fuzzy SVR based intention understanding model is proposed to deeply comprehending people's inner thoughts. To validate the proposed intention understanding model, simulation experiment and application are both designed based on the background of "drinking at a bar".

To confirm the validity of the proposed two-layer fuzzy support vector regression (TLFSVR) based intention understanding model, a comparative simulation experiment among the TLFSVR, SVR, and back propagation neural networks (BPNN) is carried out, where training/testing data are collected from questionnaire about relationship between emotion and intention to order at a bar, as well as the identification informations (i.e., age, gender, and nationality). According to simulation results, while clusters number $C = 2/3/6$, the TLFSVR obtains an accuracy of $70\%/72\%/80\%$, which is $23\%/26\%/33\%$ and $35.5\%/37.5\%/45.5\%$ higher than SVR and BPNN, respectively. Computational time is about 0.976s/0.935s/0.67s by using the proposed intention understanding model, indicating that computing speed is about 1.9/2/2.8 and 3.6/3.7/5.2 times faster than the SVR and BPNN, respectively.

According to the simulation experiment results, it can be found that the accuracy of the proposed understanding model is higher than that of the SVR and the BPNN, leading to the smaller AARD and higher correlation coefficient between the output and the actual value. Smaller AARD and higher correlation indicate that the TLFSVR based intention understanding model better matches the actual situation. The clusters number C and the overlap parameter η is two important parameter in the proposed understanding model, which guarantee the accuracy. The computational time is affected by the computational complexity of the proposal, which is explicitly expressed according to the number of data and computation scale, for example, n is the number of the training data, n_i is the number of the training data in the ith training subset. C is the number of clusters, and m is the numbers of layers in BPNN, then computational complexity of the SVR, the proposed TLFSVR, and the BPNN is $O(n^2)$, $O(\sum_{i=1}^{C} n_i^2)$, and $O(m \times n^2)$, respectively. Since computational complexity of the proposal is simplest among these three methods, the computational time is less than that of the SVR and BPNN.

The application experiment is developed to evaluate the validity and practicability of the proposed intention understanding model in the developing humans-robots interaction system, called mascot robot system (MRS) in our lab [24], where a scenario of "drinking at a bar" is performed with six salary men (played by postgraduates) and two eye robots (the bar lady robot and colleague robot). Six salary men are separated into three groups (i.e., age-different, gender-different, and nationality-different) to experience a scenario of "drinking at a bar, after that, they are requested to answer the questionnaire to evaluate the satisfaction of the bar lady's service. According to the application experiment result, the bar lady robot obtains an average accuracy of 77.8% and the average satisfaction level is about 4-satisified using the proposed TLFSVR based intention understanding model. Bar lady robot gets an proper accuracy which satisfies the customer, and this benefits from the proposed understanding model, which takes into account people in the different ages, different genders, and different countries may lead to different requirements. The results of the preliminary application also indicate that humans can transparently communicate with robots in some degree, this is because the bar lady robot can probably understand the customer requirements, and understanding is foundation of the communication, if get to know each other deeper, communication can start from heart to heart, then a transparent communication is naturally formed. If the communication is transparent, that means communication without mental disorder so that the distance between people is shorten, this is also good for communicating smoothly. Moreover, a comparative experiment between simulation and MRS are developed to discuss whether the results of preliminary application by using MRS were consistent with the results of simulation, According to the comparative result, the prediction accuracy by simulation and the correlation between them are 83.3% and 0.9608, respectively, indicating that the simulation can well reflect the change trends of the real situation in preliminary application.

Inspired by the preliminary application, for future research, the proposed intention understanding model can be developed as an ordering system in the bar for the business communication, where business could be international. Besides emotion and atmosphere, multi-robot behavior adaptation to intention is also a potential study base on the proposed intention understanding model. In some real application of human-robot interaction, intention understanding has been successfully applied in some autonomous robots, e.g., Pioneer 2DX robot [45] and ARoS [46], from which we infer the proposed TLFSVR based intention understanding model would be an effective way to autonomous robots for deeply understanding peoples' inner thoughts in human-robot interaction.

References

1. R. Kirby, J. Forlizzi, R. Simmons, Affective social robots. Robot. Automous Syst. **58**, 322–332
2. R.R. Murphy, T. Nomura, A. Billard, J.L. Burke, Human-robot interaction. IEEE Robot. Autom. Mag. **17**(2), 85–89
3. J.E. Young, J. Sung, A. Voida, E. Sharlin, T. Igarashi, H.I. Christensen, R.E. Grinter, Evaluating human-robot interaction. Int. J. Soc. Robot. **3**(1), 53–67
4. N. Mitssunaga, T. Miyashita, H. Ishiguro, K. Kogure, N. Hagita, Robovie-IV: a communication robot interacting with people daily in an office, in *Proceedings of International Conference on Intelligent Robots and System*, Beijing, China, pp. 5066–5072
5. J. Fasola, M. Mataric, Using socially assistive human-robot interaction to motivate physical exercise for older adults. Proceed. IEEE **100**(8), 2512–2526
6. N. Mitsunaga, C. Smith, T. Kanda, H. Ishiguro, N. Hagita, Adapting robot behavior for human-robot interaction. IEEE Trans. Robot. **24**(4), 911–916
7. S. Calinon, I. Sardellitti, D.G. Caldwell, Learning-based control strategy for safe human-robot interaction exploiting task and robot redundancies, in *Proceedings of IEEE/RSJ International Conference on Intelligent Robots and Systems*, Taibei, Taiwan, pp. 249–254
8. F. Broz, I. Nourbakhsh, R. Simmons, Planning for human-robot interaction in socially situated tasks. Int. J. Soc. Robot. **5**(2), 193–241
9. Z.T. Liu, L.F. Chen, F.Y. Dong, Y. Yamazaki , M. Wu, K. Hirota, Adapting multi-robot behavior to communication atmosphere in humans-robots interaction using fuzzy production rule based friend-Q learning. J. Adv. Comput. Intell. Intell. Inf. **17**(2), 291–301
10. Z.T. Liu, L.F. Chen, F.Y. Dong, Y. Yamazaki , M. Wu, K. Hirota, Multi-robot behavior adaptation to local and global communication atmosphere in humans-robots interaction. J. Multimodal User Inter. **8**(3), 289–303

11. A. Drenner, M. Janssen, A. Kottas et al., Coordination and longevity in multi-robot teams involving miniature robots. J. Intell. & Robot. Syst. **72**(2), 263–284
12. A. Geiger, M. Lauer, R. Urtasun, A generative model for 3D urban scene understanding from movable platforms, in *Proceedings of IEEE Conference on Computer Vision and Pattern Recognition*, Colorado, USA, pp. 1945–1952
13. M.J. Roberson, J. Bohg, G.L. Skantze, J. Gustafson, R. Carlson, B. Rasolzadeh, D. Kragic, Enhanced visual scene understanding through human-robot dialog, in *Proceedings of IEEE/RSJ International Conference on Intelligent Robots and Systems*, San Francisco, USA, pp. 3342–3348
14. M.V. Bergh, D. Carton, R.D. Nijs, N. Mitsou, C. Landsiedel, K. Kuehnlenz, D. Wollherr, L.V. Gool, M. Buss, Real-time 3D hand gesture interaction with a robot for understanding directions from humans, in *Proceedings of IEEE International Symposium on Robot and Human Interactive Communication*, Atlanta, USA, pp. 357–362
15. C. Mohiyeddini, R. Pauli, S. Bauer, The role of emotion in bridging the intention behaviour gap: the case of sports participation. Psychol. Sport Exercise **10**(2), 226–234
16. A. Olsson, K.N. Ochsner, The role of social cognition in emotion. Cognit.-Emot. Inter. **12**(2), 65–71
17. V.N. Vapnik, *The Nature of Statistical Learning Theory* (Springer, New York)
18. L. Bottou, V.N. Vapnik, Local learning algorithms. Neural Comput. **4**(6), 888–900
19. A. Shafi, W. Vishanth, Understanding citizens' behavioural intention in the adoption of e-government services in the state of Qatar, in *Proceedings of European Conference on Information Systems*, Verona, Italy
20. R.O. Orji, Impact of gender and nationality on acceptance of a digital library: an empirical validation of nationality based UTAUT using SEM. J. Emer. Trends Comput. Inf. Sci. **1**(2), 68–79
21. C. Holland, R. Hill, The effect of age, gender and driver status on pedestrians' intentions to cross the road in risky situations intentions to cross the road in risky situations. Accid. Anal. Prevent. **39**(2), 224–237
22. B. Keysar, The illusory transparency of intention: linguistic perspective taking in tex. Cognit. Psychol. **26**(2), 165–208
23. M. Hanheide, M. Lohse, H. Zender, Expectations, intentions, and actions in human-robot interaction. Int. J. Soc. Robot. **4**(2), 107–108
24. K. Hirota, F. Dong, Development of mascot robot system in NEDO project, in *Proceedings of International IEEE Conference on Intelligent Systems*, Varna, Bulgaria, vol. 1, pp. 38–44
25. C. Cortes, V.N. Vapnik, Support-vector networks. Mach. Learn. **20**(3), 273–297
26. H. Yu, J. Kim, Y. Kim, S. Hwang, Y.H. Lee, An efficient method for learning nonlinear ranking SVM functions. Inf. Sci. **209**, 37–48
27. Y. Tarabalka, M. Fauvel, J. Chanussot, J.A. Benediktsson, SVM- and MRF-based method for accurate classification of hyperspectral images. IEEE Geosci. Remote Sens. Lett. **7**(4), 736–740
28. M. Moavenian, H. Khorrami, A qualitative comparison of artificial neural networks and support vector machines in ECG arrhythmias classification. Expert Syst. Appl. **37**(4), 3088–3093
29. J.A.K. Suykens, J. Vandewalle, B.D. Moor, Optimal control by least squares support vector machines. Neural Netw. **14**(1), 23–25
30. S. Ismail, A. Shabri, R. Samsudin, A hybrid model of self-organizing maps (SOM) and least square support vector machine (LSSVM) for time-series forecasting. Expert Syst. Appl. **38**(8), 10574–10578
31. C. Huang, A reduced support vector machine approach for interval regression analysis. Inf. Sci. **217**, 56–64
32. J.P. Ferreira, M.M. Crisostomo, A.P. Coimbra, Adaptive PD controller modeled via support vector regression for a biped robot. IEEE Trans. Control Syst. Technol. **21**(3), 941–949
33. J.C. Dunn, A fuzzy relative of the ISODATA process and its use in detecting compact well-separated clusters. J. Cybern. **3**(3), 32–57
34. J.C. Bezdek, A convergence theorem for the fuzzy ISODATA clustering algorithms. IEEE Trans. Pattern Anal. Mach. Intell. **PAMI-2**(1), 18

35. W.M. Dong , F.S. Wong, Fuzzy weighted averages and implementation of the extension principle. Fuzzy Sets Syst. **21**(2), 183–199
36. D. Keltner, J. Haidt, Social functions of emotions at four levels of analysis. Cognit. Emotion **13**(5), 505–521
37. J.R. Spoor, J.R. Kelly, The evolutionary significance of affect in groups: communication and group bonding. Group Process. Intergroup Rel. **7**(4), 398–412
38. K.C. Klauer, J. Musch, Affective priming: findings and theories, in *Proceedings of Psychology of Evaluation*. Affective Processes in Cognition and Emotion, Lawrence Erlbaum, USA, pp. 7–49
39. A.K. Jain, R.C. Dubes, *Algorithms for Clustering Data* (Prentice Hall, Englewood Cliffs)
40. Z.T. Liu, Z. Mu, L.F. Chen et al., Emotion recognition of violin music based on strings music theory for mascot robot system, in *9th International Conference on Informatics in Control, Automation and Robotics*, Rome, Italy, pp. 5–14
41. G. Farchi, F. Fidanza, S. Giampaoli, S. Mariotti, A. Menotti, Alcohol and survival in the Italian rural cohorts of the seven countries study. Int. J. Epidemiol. **29**, 667–671
42. K. Tang, F.Y. Dong, M. Yuhki, Y. Yamazaki, T. Shibata, K. Hirota, Deep level situation understanding and its application to casual communication between robots and humans, in *Proceedings of International Conference on Informatics in Control, Automation and Robotics*, Reykjavik, Iceland, pp. 292–299
43. Z.T. Liu, M. Wu, D.Y. Li, L.F. Chen, F.Y. Dong, Y. Yamazaki, K. Hirota, Communication atmosphere in humans-robots interaction based on the concept of fuzzy atmosphere generated by emotional states of humans and robots. J. Autom. Mobile Robot. Intell. Syst. **7**(2), 52–63
44. Y. Yamazaki, Y. Hatakeyama, F.Y. Dong, K. Nomoto, K. Hirota, Fuzzy inference based mentality expression for eye robot in affinity pleasure-arousal space. J. Adv. Comput. Intell. Intell. Inf. **12**(3), 304–313
45. R. Kelley, A. Tavakkoli, C. King, M. Nicolescu, M. Nicolescu, G. Bebis, Understanding human intentions via hidden markov models in autonomous mobile robots, in *Proceedings of the 3rd ACM/IEEE International Conference on Human Robot Interaction*, Amsterdam, Netherlands, pp. 367–374
46. E. Bicho, W. Erlhagen, E. Sousa et al., The power of prediction: robots that read intentions, in *Proceedings of IEEE/RSJ International Conference on Intelligent Robots and Systems*, Vilamoura, Portugal, pp. 5458–5459

Chapter 12
Emotional Human-Robot Interaction Systems

This chapter introduces the basic structure of robot emotion system, analyzes the development function of the system according to the needs of users, and constructs a set of robot emotion interaction system based on the actual equipment. The interaction system of emotion robot constructed in this chapter provides an experimental platform for the verification of emotion recognition algorithm and emotion intention understanding model. At the same time, in order to verify the emotional intention understanding model proposed in this chapter, two reasonable scenarios are set up to realize the understanding of emotional intention in specific situations.

12.1 Introduction

The development of science and technology is making rapid changes in human society, but the goal of human is always the same. Human-robot interaction technology [1] has gradually shifted from computer-centered to human-centered, natural Human-robot interaction has become a new direction of Human-robot interaction technology development [2]. Emotional interaction [3] is the core and foundation of natural Human-robot interaction, which makes robots capable of emotional communication and more humanized by processing the emotional information in the process of Human-robot interaction [4]. The formation of big data, the innovation of theoretical algorithms, the improvement of computer computing capacity, and the evolution of network facilities are driving the development of artificial intelligence [5]. Intelligentization has become an important direction for the development of technology and industry [6]. Now, the robot revolution has entered the era of "Internet + artificial intelligence + emotion" [7], which requires robots to have the ability of emotional cognition.

L. Chen et al., *Emotion Recognition and Understanding for Emotional Human-Robot Interaction Systems*, Studies in Computational Intelligence 926,
https://doi.org/10.1007/978-3-030-61577-2_12

Intelligent emotional robot increasingly appear in people's life. Its system construction is usually based on the ordinary humanoid robot, emotional information retrieval framework, emotional understanding framework, emotional robot interaction framework eventually form a shape similar to human, inner emotional robot system with human emotions, cognitive ability. For example, Buddy, the first emotional companion family robot launched by Blue Frog Robotics, can naturally express emotions according to its interactions with family members [8]. Humanoid robot Sophia, developed by hanson robotics, is the first robot in history to obtain citizenship. It can display more than 62 facial expressions, understand language and remember interactions with humans [9]. Erica, an intelligent robot developed by professor Hiroshi Ishiguro's team at Osaka university in Japan and a research team at Kyoto university, can talk with people fluently, with sounds and expressions very similar to those of human beings. In addition, Erica once reported the news as a news anchor on a Japanese TV station in 2018 [10].

Existing emotional robot system mainly includes two aspects, a study is to let the robot to identify human emotional state and based on the interaction of feedback to adjust its behavior, this kind of research by identifying human facial expressions, voice signal, body posture and physiological signals for emotional information, and then according to the acquired user emotional response behavior guide robot, realizing Human-robot interaction [11]. For example, literature [12] proposed a companion robot system with the ability of emotional interaction. The robot USES visual and audio sensors to detect the user's basic emotions. The robot plays appropriate music according to the user's emotional state, and generates a soothing odor. The robot system can also automatically navigate and follow the user's side to accompany the user. Literature [13] developed a robot emotion recognition and rehabilitation system based on browser interface, which integrates physiological signals and facial expressions to identify users' emotional states, and then combines the technology of emotional expression to alleviate users' negative emotions or enhance their positive emotions.

Another kind of research explores the multi-modal emotional expression of robots. By adjusting the facial expression, speech synthesis mechanism, posture and movements of humanoid robots, the robots are endowed with as rich emotional expression capacity as human beings [14]. The humanoid robot ROBIN was developed in literature [15], which can express almost any facial expression of human beings and generate expressive gestures and body gestures. At the same time, it is also equipped with a voice synthesizer to realize the transformation, selection, expression, stimulation and evaluation of emotions, as well as some auxiliary functions. Literature [16] has developed an emotional robot system to accompany autistic children. By playing games and communicating with autistic children, emotional robots can identify the emotions expressed by interactors and correctly classify them.

At the same time, domestic and foreign research institutions are paying more attention to the research of affective computing and intelligent interaction. In the first half of 2018, the global robot market exceeded $27.7 billion. It is estimated that robots will rapidly penetrate the service industry from the industrial field, showing a broader market space than industrial robots [17]. However, it is very important for service robots to improve their service quality to integrate emotional factors into service robots and have emotional cognitive ability, so that they can understand human needs and realize a virtuous circle of emotional communication and satisfaction of needs [18].

Facing the urgent needs of human-machine emotional interaction system and emotional robot in domestic and foreign markets, the research of emotional robot system will surely show a new direction for the future application of natural interaction. The ultimate purpose of constructing the emotional robot system is to make the robot make appropriate behavioral response based on the understanding of the emotional state of the communicating object, so as to adapt to the constant changes of the user's emotion, and thus optimize the interactive service. Even more anthropomorphic robot, however, is also the basic no general intelligent system can close to the human level, the existing emotional robot is not short of human emotional ability, how to let the machine has more accurate emotion recognition and the intention of the deeper understanding ability, and guide the act more natural, proper response, is essential to the natural human computer interaction.

At present, some progress has been made in the research of emotional robot system related to the understanding of emotional intention and natural interaction technology. However, the research is still in its infancy, and there are still many problems to be further solved, mainly reflected in the following aspects: The research on facial expression [19], speech emotion, body posture [20], hand gestures and body language) [21], physiological signals (EEG) [22] and pulse [23] on development of recognition, but at this stage of multimodal information fusion method is not sufficiently considering the modal characteristics in identifying different emotion plays different role, under the condition of not both the correlation and differences between different modal characteristics; The research on the understanding of emotional intention is still in the preliminary stage, and the deep structure of emotional intention perception is relatively simple [24], which has not yet realized the multi-dimensional understanding of emotional intention. At the present stage, emotional models are not closely related to Human-robot interaction means [25], so it is particularly crucial to study more general and fuzzy emotional models that conform to human emotional changes and introduce them into the behavioral decision-making and coordination mechanism of robots. There are still many deficiencies in the emotional robot system, such as not considering the environmental information and scene information, which will also affect the emotional expression of the interactor, or not considering the deep intention information behind the user's emotional expression [26], and lack of deep cognitive analysis of the user's behavior, etc., which cannot form a deep social relationship.

Therefore, on the basis of the existing emotional robot system considering modal emotional information and environmental information, introducing emotional intention understanding framework, combined with the geometrical visual SLAM algorithm to build a robot autonomous positioning and visual navigation model, realize the emotional robot in emotional interaction in the process of environmental perception, so as to take advantage of new Human-robot interaction technology, build emotional robot system with emotional intention interaction ability, is particularly necessary.

In order to realize a human-robot interaction system with certain emotion recognition and intention understanding, and to establish a natural and harmonious human-robot interaction process, this section proposes a human-robot interaction system scheme based on affective computing. First, the overall architecture design and application scenario of the system are introduced, and then the application experiment based on the emotional human-robot interaction system is introduced.

12.2 Basic Emotional Human-Robot Interaction Systems

Emotional robot interaction system is the facial expression recognition, emotional intention understanding into the mobile robot, the Human-robot interaction system composed of emotion computing workstations, enables the robot to according to the surface layer of people exchange information (information such as expression) to understand the intentions, and react accordingly, to better serve users, Human-robot interaction to realize the smooth interaction.

According to the needs of People's Daily life and Human-robot interaction, the main functions of the emotional robot interaction system are analyzed as follows:

(1) Emotional recognition needs. In daily communication, people use facial expressions to judge the basic emotions of the communicator, such as sadness, anger, surprise, disgust, fear, neutrality and happiness. In Human-robot interaction, the mechanical response is difficult to meet People's Daily needs. People also hope that robots can communicate with people naturally and emotionally, and the nonverbal information conveyed by facial expressions plays an important role in People's Daily life and communication. Therefore, facial expression recognition has become a new development trend in Human-robot interaction.

(2) Users' intention to understand requirements. In daily interactions, people not only perceive the actions of others, but also understand the intention behind the actions according to the surrounding environment and the relationship between people and the actions. Understanding the intentions of others is a discipline and is considered an important part of human cognitive behavior. After understanding the user's intention, the robot can give appropriate feedback, which can improve the intimacy of Human-robot interaction.

(3) Emotional intention to express needs. The system can make the robot or virtual robot in the system give feedback or behavioral response corresponding to the result of the user's emotional intention according to the understanding of the result of the emotional intention, so as to realize the feedback of the robot to the user's emotional intention, and thus complete the whole interaction process between the robot and the robot.

(4) Multi-robot cooperation needs. As to expanding the scope of Human-robot interaction, it is anticipated that more robot systems that can cooperate with each other, on the one hand, can produce unable to realize the function of a single robot. On the other hand, fully unleashing the potential of multi-robot cooperation, improves

system efficiency, reduce consumption of energy and time, thus improving the overall performance of the system.

According to the user's demand analysis, the system needs to have the following several functional modules:

(1) The information collection module is mainly responsible for the collection of surface information. The Kinect sensor of the mobile robot can collect the user's facial expression information and then input the user's personal information in real time through the Mysql database.

(2) Feature extraction module completes the collection of various features of emotional intention. Feature collection mainly includes facial expression information and personal information. Due to the real-time performance of man-machine interaction, mainly for dynamic facial expression recognition, the application of the dynamic feature point extraction algorithm based on the Candide3 model is adopted to obtain facial expression in real time and further obtain personal information. The characteristic data of intention understanding is obtained from this.

(3) The information processing module mainly completes the processing and recognition. This module is the core part of the system and performs the main intention-understanding task. This part includes commonly used emotion recognition, emotion intention understanding model, expression of emotion intention, multi-robot cooperation.

(4) The information feedback module is responsible for giving feedback to the information of the system. This part receives feedback from the information processing module, displays the results of the execution on the screen, and drives the robot to give corresponding feedback.

The hardware topology of the emotional robot interaction system is shown in Fig. 12.1. The whole system consists of 2 mobile robots, 1 emotional computing workstation, router and data transmission equipment. The two mobile robots are connected to the wireless router via WLAN, the Kinect device is connected to the hub via USB, and the hub, the wireless router, the emotional computing workstation and the Human-robot interaction interface are connected via wire. It can be seen that

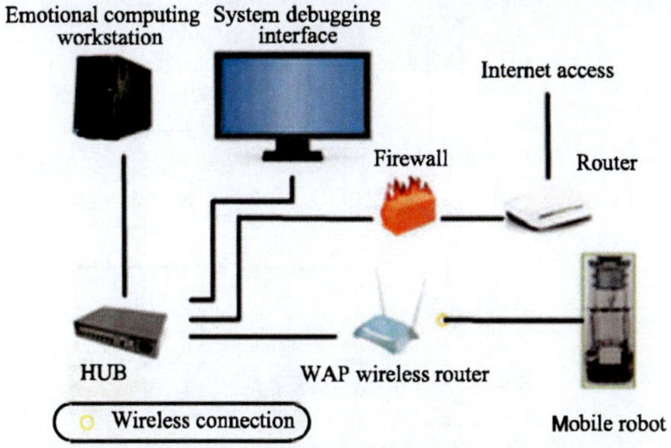

Fig. 12.1 The framework of emotional social robot system

the hardware of the system mainly includes the following three parts: Kinect sensor for facial expression acquisition, mobile robot for emotional expression and interaction, and emotion computing workstation for the test and evaluation of intention comprehension rate.

12.3 Design of Emotional Human-Robot Interaction System

The ability of robots to correctly recognize emotional states, understand emotional intentions, and give emotional feedback is an important feature and challenge in the study of emotional robot systems. Emotional robot system is a highly intelligent machine system, and the Internet can solve the problem of remote access and control of robots, as well as implement complex operations, data search, and information communication. Therefore, based on the above research content, we have constructed a natural interactive emotional robot system under the Internet architecture, enabling users to conduct natural emotional interactions with emotional robots through multiple interaction channels.

Based on the facial expression, speech emotion, body gesture, and personalized information, as well as environmental information during human-robot interaction, we analyze the correlation between human-robot interaction process data and emotional intention in specific environmental scenarios, and establish multi-dimensional emotional intention understanding model, to build an emotional robot system under the Internet, and realize real-time remote information interaction of multiple emotional robots in a specific scene.

The topological structure of the natural Internat-based interactive emotional robot system is shown in Fig. 12.2. The system structure can be divided into three layers: Web client layer, network service layer and robot local service layer. The web client

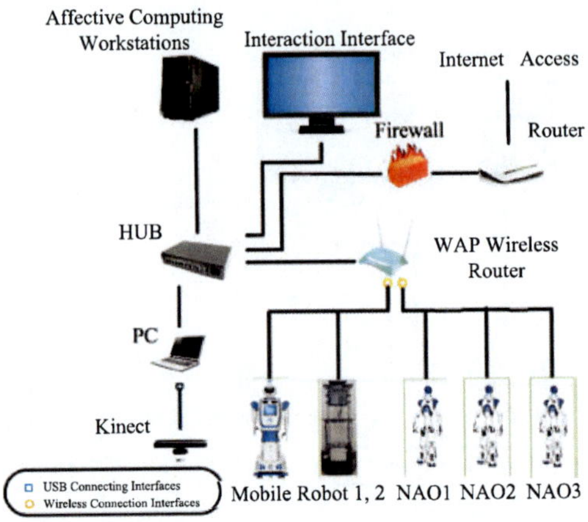

Fig. 12.2 The structure of the natural interactive emotional robot system

layer mainly provides users with an interactive control interface, and monitors the task execution and the operation of the robot equipment in real time through the human-robot interaction interface; network service layer mainly receive instructions issued by the web client layer, forward specific instructions to the local control layer of the robot, or directly call the corresponding service. At the same time, this layer also needs to accept the emotional intent-related information obtained by the local control layer of the robot and complete the corresponding computing tasks (feature extraction and recognition of emotional intention-related information, multi-modal emotional intention understanding) and feed back the calculation results (behavioral decision and coordination of affective robots) to the local control layer. The robot local service layer mainly obtain information about emotional intention and send it to the network service layer, and perform calculation results fed back from the network service layer.

The emotional robot in the system has functions of facial expression recognition, speech recognition, body gesture recognition, emotion expression, emotional intention understanding, and natural human-robot interaction. We use Kinect and wearable device to obtain facial expression, speech, and body gesture. In the Windows/Linux environment, by using languages such as Matlab, C++/C, or Python, the proposed algorithms for emotion recognition, multi-modal emotional intention understanding, and emotional robot natural interaction technology are programmed and compare with existing algorithms to verify the feasibility and superiority, and then to verify the correctness and effectiveness of the fundamental research.

12.4 Summary

This chapter introduces the basic structure of the emotional robot system, and according to the needs of users, the development of the system function analysis, combined with the actual equipment, to build a set of emotional robot interaction system. At the same time, in order to verify the emotional intention understanding model proposed in this chapter, the understanding of emotional intention under specific situations is realized by combining scenarios. The interaction system of emotion robot constructed in this chapter provides an experimental platform for the verification of emotion recognition algorithm and emotion intention understanding model.

References

1. D. Schuller, B.W. Schuller, The Age of Artificial Emotional Intelligence. Computer **51**(9), 38–46 (2018)
2. F. Li, J. Feng, M. Fu. A Natural Human-robot interaction Method in Virtual Roaming. In *Proceedings of the 2019 15th International Conference on Computational Intelligence and Security (CIS)*, 2019

3. A. Vinciarelli, M. Pantic, and H. Boulard. Social Signal Processing: Survey of an Emerging Domain. Image and Vision Computing, 2009, 27 (12): 1743-1759
4. J. C. Gómez, A. G. Serrano, P. Martl nez. Intentional processing as a key for rational behaviour through Natural Interaction. Interacting with Computers. 2006, 18 (6): 1419-1446
5. J. Fan, M. Campbell, B. Kingsbury. Artificial intelligence research at IBM. IBM Journal of Research and Development, 2011, 55 (5): 16: 1-16: 4
6. X.Q. Zheng, M. Shiomi, T. Minato, H. Ishiguro, What Kinds of Robot's Touch Will Match Expressed Emotions? IEEE Robotics and Automation Letters **5**(1), 127–134 (2019)
7. P. Christopher, L. Johnson, D. A. Carnegie. Mobile Robot Navigation Modulated by Artificial Emotions. IEEE Transactions on Systems, Man, and Cybernetics, Part B (Cybernetics), 2009, 40 (2): 469-480
8. Blue Frog Robotics. Buddy: The first emotional companion robot. http://www. Bluefrogrobotics.com/en/buddy_the-emotionnal-robot/, 2018
9. Hanson Robotics. Sophia. http://www.hansonrobotics.com/robot/sophia/, 2016
10. Live Science. Meet Erica, Japan's Next Robot News Anchor. https://www.livescience.com/61575-erica-robot-replace-japanese-news-anchor.html, 2018
11. S. B. Lee, S. H. Yoo. Design of the companion robot interaction for supporting major tasks of the elderly. In *Proceedings of the 14th International Conference on Ubiquitous Robots and Ambient Intelligence (URAI)*, 2017
12. J. H. Lui, H. Samani, K. Y. Tien. An affective mood booster robot based on emotional processing unit. In *Proceedings of 2017 International Automatic Control Conference*, 2018: 1-6
13. T. Somchanok, O. Michiko, Emotion recognition using ECG signals with local pattern description methods. International Journal of Affective Engineering **15**(2), 51–61 (2016)
14. S. Saunderson, G. Nejat, It Would Make Me Happy if You Used My Guess: Comparing Robot Persuasive Strategies in Social HumanÂ¨CRobot Interaction. IEEE Robotics and Automation Letters **4**(2), 1707–1714 (2019)
15. M. Klug, A. Zell, Emotion-based human-robot interaction. Proceedings of IEEE International Conference on Computational Cybernetics 365–368 (2013)
16. L. Boccanfuso, E. Barney, C. Foster, Emotional robot to examine differences in play patterns and affective response of children with and without ASD. Proceedings of ACM/IEEE International Conference on Human-Robot Interaction 19–26 (2016)
17. M. Ficocelli, J. Terao, G. Nejat, Promoting interactions between humans and robots using robotic emotional behavior. IEEE Transactions on Cybernetics **46**(12), 2911–2923 (2016)
18. F. Xu, J. Zhang, J. Wang, Microexpression identification and categorization using a facial dynamics map. IEEE Transactions on Affective Computing **8**(2), 254–267 (2017)
19. M. Sheikhan, M. Bejani, D. Gharavian, Modular neural-SVM scheme for speech emotion recognition using ANOVA feature selection method. Neural Computing and Applications **23**(1), 215–227 (2013)
20. Z. Yang, S. Narayanan, Modeling dynamics of expressive body gestures in dyadic interactions. IEEE Transactions on Affective Computing **8**(3), 369–381 (2017)
21. Y. Ding, X. Hu, Z. Y. Xia, Y. J. Liu, D. Zhang. Inter-brain EEG feature extraction and analysis for continuous implicit emotion tagging during video watching. IEEE Transactions on Affective Computing. https://doi.org/10.1109/TAFFC.2018.2849758.
22. J.C. Peng, Z. Lina, Research of wave filter method of human sphygmus signal. Laser Technology **40**(1), 42–46 (2016)
23. C.L.P. Chen, Z. Liu, Broad learning system: an effective and efficient incremental learning system without the need for deep architecture. IEEE Transactions on Neural Networks and Learning Systems **29**(1), 10–24 (2018)
24. Z. T. Liu, F. Y. Dong, K. Hirota, M. Wu, D. Y. Li, Y. Yamazaki. Emotional states based 3-D fuzzy atmosfield for casual communication between humans and robots. In *Proceedings of International Conference on Fuzzy Systems*, Taipei, 2011: 777-782
25. S.Y. Jiang, C.Y. Lin, K.T. Huang, Shared control design of a walking-assistant robot. IEEE Transactions on Control Systems Technology **25**(6), 2143–2150 (2017)
26. X. Lu, W. Bao, S. Wang, Three-dimensional interfacial stress decoupling method for rehabilitation therapy robot. IEEE Transactions on Industrial Electronics **64**(5), 3970–3977 (2017)

Chapter 13
Experiments and Applications of Emotional Human-Robot Interaction Systems

The ability of robots to correctly recognize emotional states, understand emotional intentions, and give emotional feedback is an important aspect and challenge in the study of emotional robot systems. In order to realize a human-robot interaction system with certain emotion recognition and intention understanding, and to establish a natural and harmonious human-robot interaction process, this section proposes a human-robot interaction system scheme based on affective computing. First, the overall architecture design and application scenario of the system are introduced, and then the application experiment based on the emotional human-robot interaction system is introduced.

13.1 Introduction

The fundamental starting point of emotional computing research is that people no longer are satisfied the current passive Human-robot interaction that relies on the traditional, but remote from being a more harmonious and natural Human-robot interaction [1]. When the machine has emotional intelligence, it will be able to accurately identify human emotions, and adapt to the atmosphere to adapt and pre-judgment the user's intention to achieve the purpose of active service, Human-robot interaction will be promoted to a new height [2]. Robots can correctly generate and express emotional states, which is an important content and challenge of emotional robot research [3].

In order to communicate with people naturally, the robot should be able to both understand the emotions of the user and express its emotions [4, 5]. This requires the robot to perceive the user's facial expressions [6] and posture information [7] in real time through the visual device; the user's speech, voice posture and semantic

L. Chen et al., *Emotion Recognition and Understanding for Emotional Human-Robot Interaction Systems*, Studies in Computational Intelligence 926,
https://doi.org/10.1007/978-3-030-61577-2_13

information [8] in real time through the voice acquisition device; further, with the development of various wearable sensing devices, you can pass the human physiological signal acquisition device senses the user's EEG [9], EMG [10] and other physiological signals [11] in real time, thereby more accurately analyzing the user's emotional state [12].

At present, many research institutions in the world are actively researching emotional information processing technologies (expression and speech emotion recognition, wearable device development, intent understanding methods, atmosphere adaptation technology, etc.) [13]. Currently well-known emotional intelligence research units include: the emotional speech group led by Cowie and Douglas-Cowie of Queen's University of Belfast; the media research laboratory of MIT; the human-machine speech interaction group in charge of Schuller of Munich University of Technology; Southern California The speech emotion group in charge of the university Narayanan; the emotion research laboratory led by Soberer of the University of Geneva; the emotion robot research group led by Canamero of the Free University of Brussels, etc. [14]. Based on the inductive analysis of the research results of these well-known institutions, this section selects two types of early and more mature Human-robot interaction systems: Mascot human-computer emotional interaction system [15], WE-4RII human-computer emotional interaction system [16], from the design concept Introduce to the framework construction, and summarize the typical applications of emotional robots based on this.

In order to realize a human-robot interaction system with certain emotion recognition and intention understanding, and to establish a natural and harmonious human-robot interaction process, this section proposes a human-robot interaction system scheme based on affective computing [17]. First, the overall architecture design and application scenario of the system are introduced, and then the application experiment based on the emotional human-robot interaction system is introduced.

The ability of robots to correctly recognize emotional states, understand emotional intentions, and give emotional feedback is an important content and challenge in the study of emotional robot systems [18, 19]. Emotional robot system is a highly intelligent machine system, and the Internet can solve the problem of remote access and control of robots, and implement complex operations, data search, and information communication [20–22]. Therefore, based on the above research content, we have constructed a natural interactive emotional robot system under the Internet architecture, enabling users to conduct natural emotional interactions with emotional robots through multiple interaction channels [23].

Based on the facial expression, speech emotion, body gesture, and personalized information, as well as environmental information [24, 25] during human-robot interaction, we analyze the correlation between human-robot interaction process data and emotional intention in specific environmental scenarios, and establish multi-dimensional emotional intention understanding model, to build an emotional robot system under the Internet, and realize real-time remote information interaction of multiple emotional robots in a specific scene [26].

The whole system will use emotional computing, multi-modal information fusion, Human-robot interaction technology, through multiple channels of sensors, robots and other equipment to build an emotional Human-robot interaction system experimental platform [27]. The system structure can be divided into three layers: Web client layer, network service layer and robot local service layer. The web client layer mainly provides users with an interactive control interface, and monitors the task execution and the operation of the robot equipment in real time through the human-robot interaction interface; network service layer mainly receive instructions issued by the web client layer, forward specific instructions to the local control layer of the robot, or directly call the corresponding service [28]. At the same time, this layer also needs to accept the emotional intent-related information obtained by the local control layer of the robot and complete the corresponding computing tasks (feature extraction and recognition of emotional intention-related information, multi-modal emotional intention understanding) and feed back the calculation results (behavioral decision and coordination of affective robots) to the local control layer. The robot local service layer mainly obtains information about emotional intention and send it to the network service layer, and perform calculation results fed back from the network service layer.

The emotional robot in the system has functions of facial expression recognition, speech recognition, body gesture recognition, emotion expression, emotional intention understanding, and natural human-robot interaction. We use Kinect and wearable device to obtain facial expression, speech, and body gesture. In the Windows/Linux environment, by using languages such as Matlab, C++/C, or Python, the proposed algorithms for emotion recognition, multi-modal emotional intention understanding, and emotional robot natural interaction technology are programmed and compare with existing algorithms to verify the feasibility and superiority, and then to verify the correctness and effectiveness of the theoretical research.

13.2 Emotional Interaction Scenario Setting

The system takes the school teaching scene and open discussion scene, as shown in Figs. 13.1 and 13.2, constructs the human-robot interaction process, and explores the application of the natural interactive emotional robot system under the Internet.

In the school teaching scene, the emotional robot can identify the teachers and students' emotion and intention information to obtain the students' degree of knowledge understanding during the lecture, provide reference for the optimization of teaching mode, and make the teaching more intelligent. At the same time, the emotional robot can also play an active atmosphere in the classroom, helping the teaching process to progress smoothly. In an open discussion scenario in the lab, an emotional robot can take notes of a meeting by identifying what attendees are saying. At the same time, the participants' emotional intention can be analyzed through their facial expressions, voice, body posture, physiological signals and personalized information, etc., and corresponding emotional expressions can be generated to regulate the atmosphere of the meeting and provide certain services. Natural human-robot interaction system is divided into emotional intention related information acquisition,

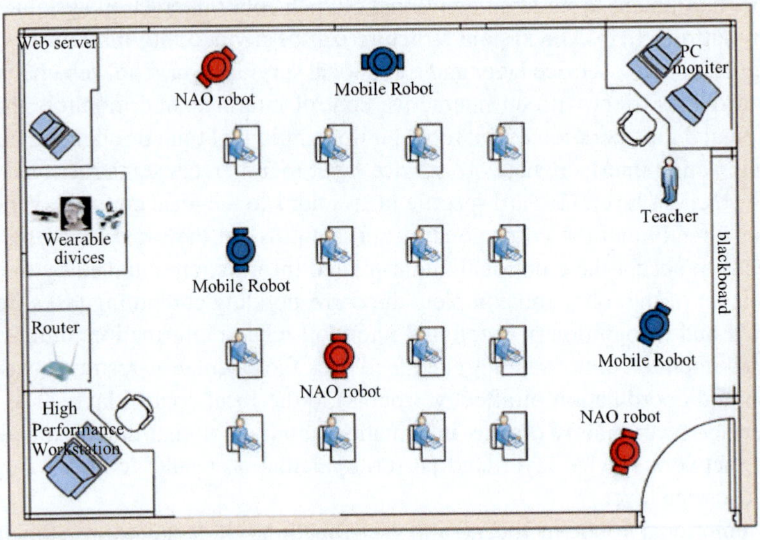

Fig. 13.1 The school teaching scene

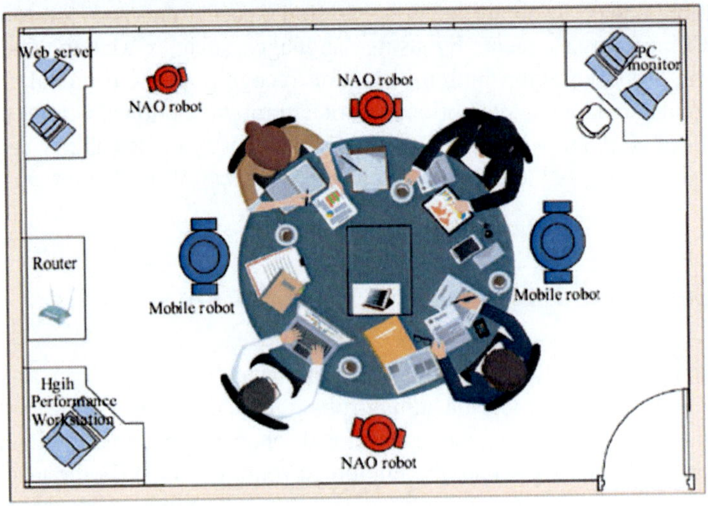

Fig. 13.2 The open discussion scene

multi-dimensional emotional intention understanding and natural interactive behavior decision-making and coordination. In the acquisition of communication information, the emotional robots obtain communication information through sensing devices, including facial information, voice, body gesture, and personalized information. Each emotional robot transmits the communication information to the emotional intention understanding part via Ethernet/wireless network. The high-performance computer in the part of emotional intention understanding perform complex computing on the multidimensional emotional intention model based on the analysis of emotional intention-related information. The goal of behavior decision-making and coordination is mainly to determine the robot's behavior based on the current emotional intention. Based on the calculation of the reward and punishment function, the high-performance computer performs robot behavior decision-making and coordination, and sends the emotion to the execution mechanism of each emotional robot in real time.

13.3 Multi-modal Emotion Recognition Experiments Based on Facial Expression, Speech and Gesture

In this section, we apply the proposed emotional recognition algorithm to our emotional robot system, and mainly introduce the research results of softmax regression based deep sparse autoencoder network for facial emotion recognition in emotional robot system.

The network of emotional robot system includes two wheel robots, three NAO robots, a Kinect sensor, and a high-performance computer for affective computing. Both NAO robots and mobile robots are connected to the wireless router via wifi. Eye tracker, Kinect, and wearable sensing devices access to a PC that is responsible for capturing emotional information and controlling devices via USB interface and wifi. The PC and wireless router are connected to the server and workstation via a hub. NAO robots can capture video images and audio data for emotion recognition of humans. In turn, NAO robots can express its own emotions by using speech, body gesture, and movement according to the humans' emotion and intension. Mobile robot can move freely, which consists of an IPC, a Kinect, and a 16∗32 LED screen. The speech synthesis software and microphone in IPC can be used for emotional expression of the mobile robot. The Kinect is a 3D somatosensory camera that can capture dynamic image and conduct image recognition and voice recognition. LED screen display can be used to express mobile robot's basic emotional expression.

System workflow uses kinect that on the top of wheeled robot to track facial expression images firstly, then invokes facial emotion recognition algorithm to feature extraction and emotion recognition, which is designed by MATLAB, and relies on the affective computing workstations as shown in Fig. 13.3.

Moreover, JAFFE facial expression database is used as training sample, which contains 213 facial expression images, including ten subjects and seven types of basic expressions. The sample images of JAFFE are shown in Fig. 13.4.

Visualization of underlying characteristics for weights that learned by sparse autoencoder network is designed, and the numbers of neurons node in the hidden layer are set to 140 to obtain an initial image of characteristics' visualization. Visual characteristics weight matrix is shown as follows.

Figures 13.5 and 13.6 are characteristic matrix of weights visualization images before and after fine-tune overall weights, it can be seen that the features self-learned

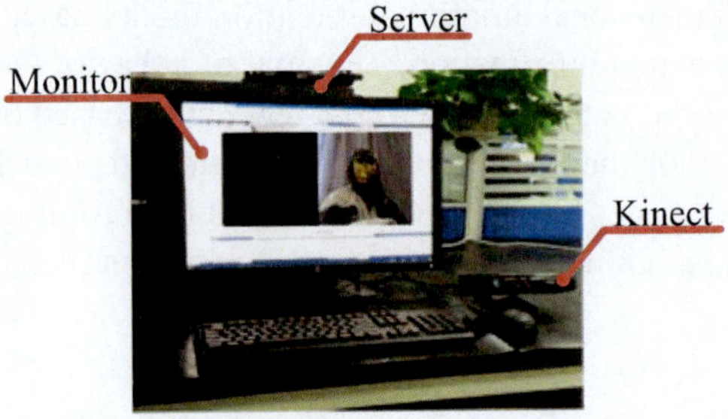

Fig. 13.3 Emotional computing workstation

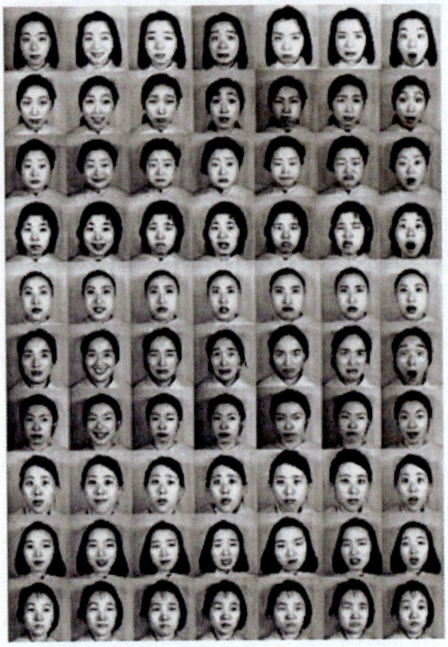

Fig. 13.4 Samples of JAFFE facial expression

Fig. 13.5 The weights visualization of the underlying characteristics

Fig. 13.6 The fine-tuning the weights visualization of the underlying characteristics

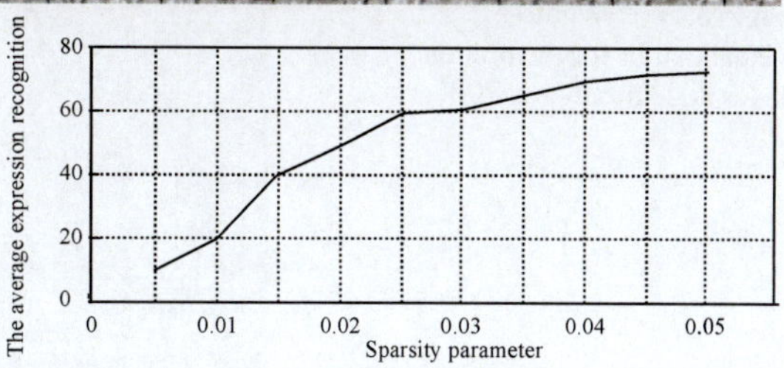

Fig. 13.7 The influence of sparse parameter and fine-tuning the weights on the rate of facial emotion recognition

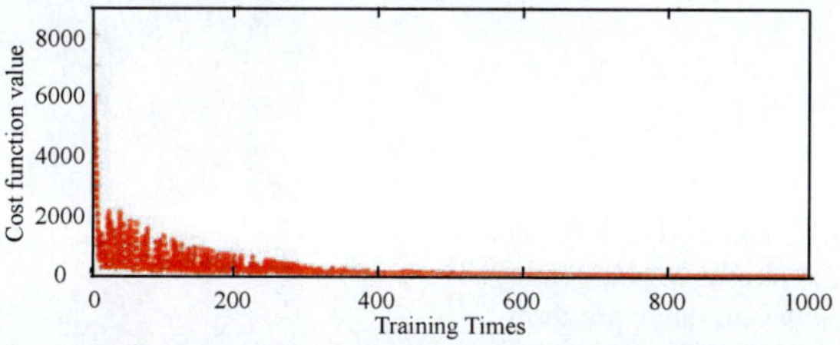

Fig. 13.8 Convergence of overall cost function before fine-tuning

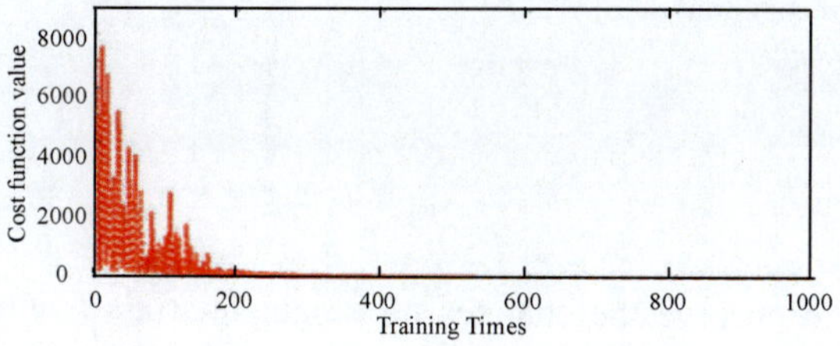

Fig. 13.9 Convergence of overall cost function after fine-tuning

by overall network look more sophisticated after fine-tune the weights to ensure high recognition accuracy. The relation curve between the seven kinds of facial emotion recognition rate and the sparsity parameter is shown in Fig. 13.7.

Comparing results shown in Fig. 13.8 with Fig. 13.9, fine-tune can make the overall cost function converge faster, and in the actual test of 182 times when they have converged and stopped training.

Design the visualization of underlying characteristic for weights in the sparse autoencoder network, and change the numbers of neurons node in the hidden layer

Fig. 13.10 Weights visualization for the hidden layer node number is 50

Fig. 13.11 Weights visualization for the hidden layer node number is 200

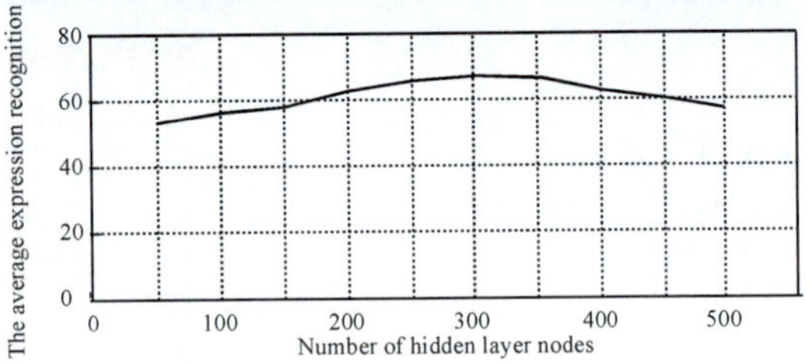

Fig. 13.12 The influence of the number of hidden layer nodes on the expression recognition rate

to observe the changes of weight matrix's feature images. Visual characteristics weight matrix is shown as follows.

It can be seen from Figs. 13.10 and 13.11 that increase the numbers of hidden layer nodes can increase the recognition rate of emotion, however, too large numbers of hidden layer nodes is helpless to improve the recognition rate of emotion. It will cause network over fitting and not conducive to expression feature extraction. After fine-tune the weight matrix, the recognition rate will be improved to a certain extent, and offset the impact of the hidden layer nodes numbers change, shown in Fig. 13.12.

To verify the accuracy of emotion recognition, simulation experiments are developed by using MATLAB, and the results are listed in Table 13.1.

Table 13.1 Comparison of emotion recognition experimental results

Emotion	Softmax regression (%)	SRDSAN (%)
Natural	86.667	100.00
Happy	80.000	93.333
Sad	63.333	100.00
Fear	63.333	76.667
Angry	83.333	100.00
Disgust	66.667	93.333
Surprise	70.000	100.00
Average	73.333	94.761

According to Table 13.1, the average recognition accuracy of Softmax regression is 73.33% in the final test. Nevertheless, if we use unlabeled training data to train deep autoencoder network firstly, and then train the Softmax regression model, and we can find that the times of iteration convergence are only 181, and the accuracy is 94.76%. By fusing the Softmax retrogression in deep learning, it can be seen that the features self-learned by overall network look more sophisticated after fine-tune, and fine-tune makes the overall cost function converge faster, which overcomes the local extrema and gradient diffusion problems. Moreover, this shows that the characteristics learned by our self-learning sparse autoencoder network are more representative than the characteristics in original input data, which is the typical difference between conventional training methods and deep learning training methods. In addition, texture and shape feature changes in the three key parts (region of interest, ROI) such as eyebrows, eyes, mouth may reflect changes in facial expression features exactly (Fig. 13.13).

The proposed Softmax-regression-based deep sparse autoencoder network will be applied to the preliminary application experiments by using the developing ESRS, as shown in Fig. 13.15. It is really an important way to autonomous robots for emotion recognition in human-robot interaction. First, the Kinect is used for image acquisition, then, the image was conducted on ROI clipping. With that, the image information is applied to the proposed method that is Softmax-regression-based deep sparse autoencoder network. In the end, facial expression results are presented as shown in Fig. 13.14.

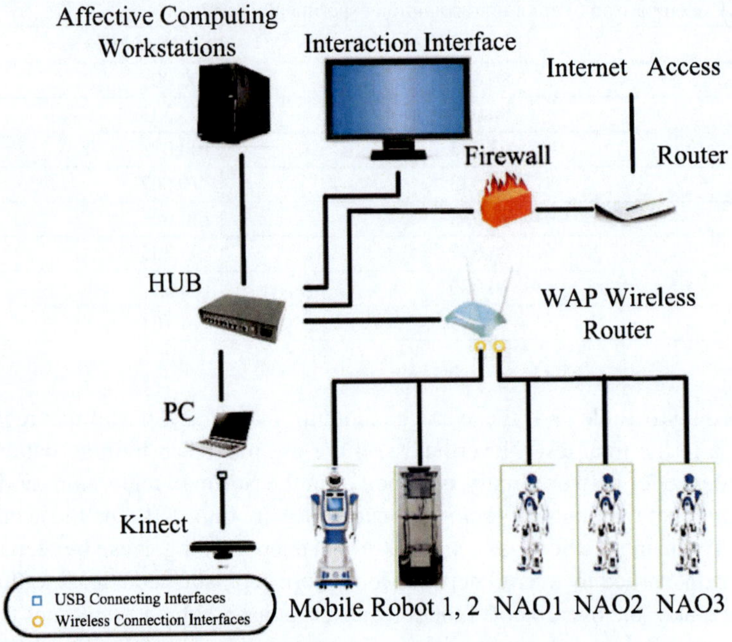

Fig. 13.13 Emotional social robot system

Fig. 13.14 The preliminary application experiments

13.4 Emotional Intention Understanding Experiments Based on Emotion, Age, Gender, and Nationality

Using the developing mascot robot system (MRS) in our lab, the proposed intention understanding model based on the emotion is evaluated by performing a scenario of "drinking at a bar" with salary men (played by postgraduates) and two eye robots (i.e., bar lady robot and colleague robot), as shown in Fig. 13.16. In the scenario of

Fig. 13.15 The preliminary application experiments

"drinking at a bar", bar lady robot needs to understanding the customers' intentions to order based on the emotions. There groups of volunteers are invited to experience the scenario, which are age-different group (i.e., two male aged 24 and 33 both from China), gender-different group (i.e., one male and one female both aged 25 from Japan), and nationality-different group (i.e., one male from Japan and the other from China both age 28).

In the experiment, six primary emotions mentioned, i.e., happiness, surprise, fear, anger, disgust, sadness are used to represent human emotions. In the dialogue between humans and robots, emotional keywords and phrases are used for expressing human emotions, such as "good job" and "well done" for happiness; "really" and "oh my god" for surprise; "worry about" and "work hard" for fear; "shut up" and "keep away from me" for anger; "bad" and "hate" for disgust; "mistake" and "stress" for sadness. In addition, emotional facial expressions by using facial movement are necessary. Moreover, several gestures are performed, such as "toast", "nod", "victory pose", "farewell", and "handclap" for happiness; "squatting with hands over head" for sadness; "face covering" for fear.

Fig. 13.16 Scenario of "drinking at a bar"

Table 13.2 Questionnaires for customer feedback

Questions	Answers
1. Is this (the drink given by bar lady robot) you want in your mind?	1-Yes, 2-no
2. How much is your satisfaction?	1-Very dissatisfied, 2-dissatisfied, 3-normal, 4-satisfied, 5-very satisfied

Instead of saying what would like to drink, each customer expresses the emotional behavior to prompt the bar lady robot. The emotional behavior includes the emotional speech facial expression, and gestures, where the interaction between customers and robots requires clear speech and standard facial expressions as well as gestures, or else it will be miss recognized.

To verify the proposal, three situations are applied by using MRS, including bar lady robot served drinks by the proposed TLFSVR based intention understanding model, SVR based intention understanding model, and randomly. After the experience the scenario of "drinking at a bar", the identification information and emotion-intention information are obtain. At the same time, the volunteer of each groups needs to answer the questionnaire to feedback, as shown in Table 13.2.

Table 13.3 shows the response of volunteers after experience the three situations. To show the statistical significations of the response, a one-way ANOVA is used to compare the means of different accuracy (intention understanding)/satisfaction in the three different situations to determine whether they are significantly different from each other. The null hypothesis is all the three situations have equal influence on intention understanding accuracy and satisfaction. By using one-way ANOVA, boxplots of response by situations are shown in Fig. 13.17, respectively. According to ANOVA summary in Table 13.4, all the P-values are lower than 0.05, thus the null

Table 13.3 Response of questionnaires

Group	ID	Bar lady served drinks by TLFSVR/SVR/random in MRS	
		Responses for accuracy	Responses for satisfaction
Age-different	Salary man A	1/1/1	4/4/4
	Salary man B	2/1/2	3/3/2
	Colleague robot	1/2/2	5/1/1
Gender-different	Salary man A	1/2/2	4/3/2
	Salary man B	1/1/2	4/4/3
	Colleague robot	2/2/2	1/1/1
Nationality-different	Salary man A	1/1/2	4/3/2
	Salary man B	1/1/1	4/3/3
	Colleague robot	1/1/2	5/5/1

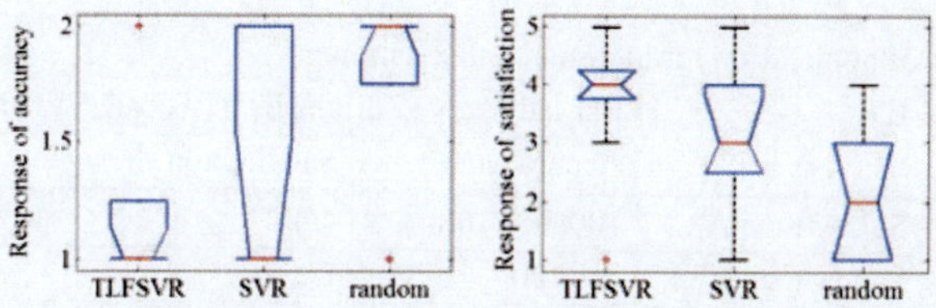

(a) Boxplots of accuracy by situations (b) Boxplots of satisfaction by situations

Fig. 13.17 Boxplots of responses by situations

hypothesis is rejected, that means different situations have signification influence on intention understanding accuracy and satisfaction.

According to the result of questionnaires in Tables 13.3 and 13.5 shows the understanding accuracy and satisfaction of each group, where response of question 1 is used to calculate the understanding accuracy of the customers' intention to order, answer "Yes" for 100% and "No" for 0%. For a robot, if the accuracy is satisfaction of the bar lady is 5-very satisfied, otherwise 1-very dissatisfied. For human, the satisfaction is based on the understanding accuracy and the service attitude (subjective evaluation by human). According to Table 13.5, using the proposed TLFSVR based intention understanding model, the bar lady robot obtains an average accuracy of 77.8% which is 11% and 50% higher than that of SVR model and served randomly, respectively; the average satisfaction level is about 4-satisfied which is one level and two levels higher than that of SVR model and served randomly, respectively.

To discuss whether the results of preliminary application by using MRS were consistent with the results of simulation, a comparative experiment are done. In the simulation, training data is the same as before as shown in Fig. 11.13, and testing data is obtain from MRS as shown in Table 11.12. According to the comparative result in Fig. 13.18 and data analysis in Table 13.6, the prediction accuracy is 83.3%, that means the simulation agrees well with the preliminary application, and the correlation between them are 0.9608, indicating that the simulation can well reflect the change trends of the real situation in preliminary application.

The results of the preliminary application indicate that humans can transparently communicate with robots to some degree. This is because the bar lady robot can probably understand the customer requirements, and understanding is a foundation of the communication, if get to know each other deeper, communication can start

Table 13.4 ANOVA summary of (a)/(b)

Source	SS	df	MS	F	$P_{rob} > F$
Columns	1.55556/12.5185	2/2	0.77778/6.25926	3.65/4.36	0.0412/0.0242
Error	5.11111/34.4444	24/24	0.21296/1.43519		
Total	6.66667/46.963	26/26			

Table 13.5 Result of application for intention understanding

Group	ID	Bar lady served drinks by TLFSVR/SVR/random in MRS	
		Accuracy (%)	Satisfaction
Age-different	Salary man A	100/100/100	4/4/4
	Salary man B	0/100/0	3/3/2
	Colleague robot	100/0/0	5/1/1
Gender-different	Salary man A	100/0/0	4/3/2
	Salary man B	100/100/0	4/4/3
	Colleague robot	0/0/0	1/1/1
Nationality-different	Salary man A	100/100/0	4/3/2
	Salary man B	100/100/100	4/3/3
	Colleague robot	100/100/0	5/5/1
		77.8/66.7/22.2 (Average)	4/3/2 (Average)

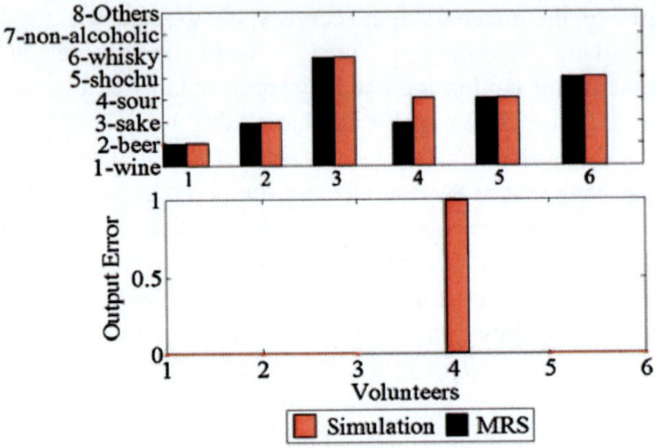

Fig. 13.18 Comparative experiment between simulation and MRS

Table 13.6 Result analysis of Comparative experiment

Prediction accuracy (%)	Correlation
83.3	0.9608

from heart to heart, then a transparent communication is naturally formed. Transparent communication becomes more and more important in our daily life, especially for building trust from each other, for example, the application at a bar using the proposal, the bar lady robot try to understand the customer, actually they are building trust mechanism during the understanding process, and the customers feedback high satisfaction is improvement for the communication. If the communication is transparent, that means communication without mental disorder in such a way that the distance between people is shorten, this is also a boost for communicating smoothly.

In the application experiment, bar lady robot gets a proper accuracy which satisfies the customer, and this benefits from the proposed intention understanding model, which takes into account people in the different ages, different genders, and different countries may has their own culture, resulted in different requirements. Reference from the preliminary application, the proposed intention understanding model can be extended to an ordering system in the bar for the business communication, where the business men are from different countries with different ages, different genders, and different culture backgrounds. Restaurant is similar to the bar, both have order system, based on the proposed intention understanding model, besides age, gender, nationality, more identification information are needed to add for the restaurant's order system, such as dietary opinions and food taboos. In addition, since people may has different intentions in different scenarios, a general intention understanding model is not easy to build. But in the proposed model, age, gender, and nationality are used as the general information, even other information can be added, which indicates the proposal could be a general model for the intention understanding, and one of the possible extension is group discussion, another important environmental information—communication atmosphere should be considerate in humans-robots interaction according to our previous research in such a way that it would an interesting topic to extent the application of the proposal to more general scenario by fusing other identification or environmental information.

Moreover, deep level understanding is being popular in human-robot interaction, and research topics would be interesting to extent the previous multi-robot behavior adaptation mechanisms from emotion and atmosphere to intention. As the development of the cognitive science, in the real application of human-robot interaction, intention understanding has been successfully applied in some autonomous robots, e.g., Pioneer 2DX robot understands people's walking direction from experience using the Hidden Markov Models and the ARoS (ActivMedia Robotics Operating System) robot as sociable partners in collaborative joint activity based on cognitively understanding of people's actions, showing the evidence that the proposed TLFSVR based intention understanding model may be an effective way to autonomous robots for deeply understanding peoples' inner thoughts in human-robot interaction.

13.5 Application of Multi-modal Emotional Intention Understanding

The proposed intention understanding model is verified by using the developing emotional social robot system (ESRS) in our lab, where two emotional robots (information robot and deep cognitive robot) and thirty customers (volunteers from postgraduates) experience the scenario of "drinking at a bar". ESRS is shown in Fig. 13.19, which is consist of information sensing and transfer layer and intelligent computation layer.

13.5.1 Self-built Data

30 volunteers aged 18–28 are invited to build up the experimental data, and they are postgraduates of our lab or undergraduate students. All of them come from China from different provinces, and the influence of the province is also taken into account. Each volunteer expresses seven primary emotions, i.e., happiness, anger, sadness, disgust, surprise, fear, and neutral, and human emotional intention includes the 8 kinds such as 1-wine, 2-beer, 3-sake, 4-sour, 5-shochu, 6-whisky, 7-non-alcoholic drink, and 8-others (other drinks or food). The desire for each drink is divided into

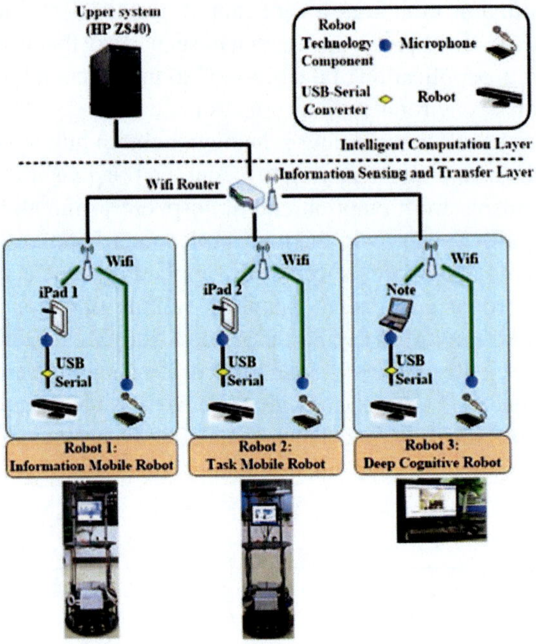

Fig. 13.19 Network of ESRS

5 levels, i.e., very weak (1), weak (2), neutral (3), strong (4), and very strong (5). Volunteers can choose 1 to 5 in the questionnaire to express the desire for each drink in different emotion. For information mobile robot, candide3 based dynamic feature point matching method is used to recognize the dynamic emotion; for deep cognitive robot, human emotional intention is understood by using proposed TLFSVR-TS model. In fact, 420 groups of data are collected in total, where half for training and other half for testing.

13.5.2 Experiments on Dynamic Emotion Recognition and Understanding

Dynamic emotion recognition and emotion understanding are both done in the preliminary application. According to the real-time dynamic facial expression of humans, 210 dynamic emotions of 30 volunteers are recognized by using the feature point matching and the introduced Candide3 based dynamic feature points matching. Experiment results are shown in Table 13.7, and confusion matrices of emotion recognition are shown in Tables 13.8 and 13.9.

According to result analyses of average recognition rate and confusion matrices, the average recognition rate is 80.48% by using the proposal, which is 10.48% higher than that of feature point matching. Due to the sliding window mechanism, which is used to select the 20 sets of the dynamic emotion for repeated confirming the emotion, the recognition rate is high than the other method. Moreover, the proposal takes into

Table 13.7 Comparison of dynamic emotion recognition experimental results

Algorithm	Average recognition rate (%)
Feature point matching	70
Candide3 based dynamic feature point matching	80.48

Table 13.8 Confusion matrices by using feature point matching

Emotion (%)	Sadness	Angry	Surprise	Disgust	Fear	Neutral	Happiness
Sadness	43.34	0	0	33.33	0	23.33	0
Angry	0	43.34	0	33.33	23.33	0	0
Surprise	0	6.67	86.66	6.67	0	0	0
Disgust	0	20.00	0	66.67	13.33	0	0
Fear	0	20.00	0	16.67	63.33	0	0
Neutral	0	0	0	0	0	100	0
Happiness	0	0	0	0	0	13.33	86.67

Table 13.9 Confusion matrices by using Candide3 based dynamic feature point matching

Emotion (%)	Sadness	Angry	Surprise	Disgust	Fear	Neutral	Happiness
Sadness	66.67	0	0	23.33	0	10.00	0
Angry	0	63.34	0	23.33	13.33	0	0
Surprise	0	3.33	93.34	3.33	0	0	0
Disgust	0	13.33	0	76.67	10.00	0	0
Fear	0	10.00	0	10.00	80.00	0	0
Neutral	0	0	0	0	0	100	0
Happiness	0	0	0	0	0	16.67	83.33

account correlations of feature points in each facial action units (AU), which can find out the variation of main feature points in facial expression, in such a way that improving emotion recognition rate. Since emotion expression is varied person to person, the proposal can handle such problem in two ways. One is that database of feature points corresponding to each emotion are collected from volunteers (consist of male and female). In the case of a stranger, we can match the real-time data to the database. Because different genders has their own feature, the matching can be divided into two clusters, i.e., male and female, and in such a way that the recognition rate can be improved. The other is that the fuzzy rules for emotion recognition are established. In the case of no full matching of the feature points, fuzzy rules that match most of the feature points will be selected, which guarantees the recognition rate to some extent.

To show the efficiency of the proposed emotion understanding model, comparative experiments are carried out among the TLFSVR-TS, the TLFSVR, KFCM-FSVR, and the SVR, which aims to understand the intention and discuss the impact of gender/province/age on the emotional intention understanding. Furthermore, to establish TS fuzzy model, the chapter analyzes thirty volunteers' intention to drink in the home environment. According to data analysis, we found that preferences of men and women are different in the home environment. Moreover, it is not difficult to find that most people would like to drink beer and shochu when they feel displeasure. But in turn, they prefer non-alcoholic drink or foods. This maybe relate to the Chinese eating habits. As a result, this can serve as reference of drinking habits in the home environment.

In the experiments, inputs are emotion, gender, province, and age, output is the intention. Clusters for FCM and membership function of TS fuzzy model are consider for developing the experiments. According to the gender (i.e., female and male), the number of clusters C for FCM is set as 2; considering people from different provinces in China, i.e., Hubei province (location of our lab) and non-Hubei province, the number of clusters C for the FCM is set as 2; since postgraduates (volunteers) can be divided as senior, junior, and freshman according to age, the number of clusters C is set as 3.

Table 13.10 Comparison of Emotion Understanding Experimental Results

Index	MF	TLFSVR-TS			TLFSVR			KFCM-FSVR			SVR
		Gender	Province	Age	Gender	Province	Age	Gender	Province	Age	
		$C = 2$	$C = 2$	$C = 3$	$C = 2$	$C = 2$	$C = 3$	$C = 2$	$C = 2$	$C = 3$	
SD	MAX	1.4130	1.4220	1.4511	1.4042	1.4249	1.4798	1.3971	1.4034	1.4739	1.4667
	MEAN	1.5123	1.5159	1.5377							
	MEDIAN-MAX	1.3974	1.4630	1.5264							
CC	MAX	0.6615	0.6561	0.6378	0.6710	0.6698	0.6580	0.6680	0.6560	0.6190	0.6278
	MEAN	0.6414	0.6361	0.6204							
	MEDIAN-MAX	0.6751	0.6379	0.6008							
UA (%)	MAX	76.67	76.19	75.71	70.00	70.00	69.05	70.48	69.05	68.10	67.62
	MEAN	76.19	75.71	75.24							
	MEDIAN-MAX	77.62	76.19	75.71							

Based on the experimental experience in our previous study, parameters of clustering methods are obtained by trial and error. For parameters setting of FCM, the overlap parameter η can be chose as 2.5, 3.5, 4, and 5 according to input dimension with 1, 2, 4, and 8, respectively. With this in mind, since the inputs are 4 in the experiments, the η and the sensitivity threshold ϵ are set to 3 and 1, respectively. In the SVR, width σt of Gaussian kernel is recommended as (0.1–0.5) $\times$ (rang of the inputs), and according to the inputs in the experiments are all normalized, σ can be selected as 0.5, and the scalar regularization parameter γ is set to 200.

The performance of the proposed TLFSVR-TS, TLFSVR, KFCM-FSVR and SVR with $C = 2/2/3$ in different membership functions (MF), respectively, and the results analysis are showed in Table 13.10. In the experiments, standard deviation (SD), correlation coefficient (CC), understanding accuracy (UA) are used as evaluation indexes. According to experiment results, in the case of MAX membership function, TLFSVR-TS model obtains understanding accuracy of 76.67%, 76.19%, 75.71%, while $C = 2/2/3$. In the case of MEAN membership function, TLFSVR-TS model obtains understanding accuracy of 76.19%, 75.71%, 75.24%, while $C = 2/2/3$. In the case of MEDIAN-MAX membership function, TLFSVR-TS model obtains understanding accuracy of 77.67%, 76.19%, 75.71%, while $C = 2/2/3$. The results demonstrate that the proposed TLFSVR-TS model obtains high accuracy than those of TLFSVR, KFCM-FSVR, and SVR.

According to results of the experiments, the proposed emotion understanding model receives higher accuracy than that of the TLFSVR, the KFCM-FSVR, and the SVR, leading to higher correlation coefficient and smaller squared error between the actual values and the output values. Higher correlation and smaller squared error means that our proposal can better reflect the actual situation. At the same time, by changing the clusters number C according to the identification information (i.e., gender, province, and age) and membership function (MF) based on data collection of volunteers, appropriate C and MF can be found to enhance the understanding accuracy to a certain extent. Additionally, supplementary information such as identification information, are applied to minimize the influence of missing or false emotion recognition, and in such a way that ensure the accuracy of emotion understanding.

For further research in human-robot interaction, deep level information understanding is being more and more popular, how robots adapt their behaviors to deep level information such as intention, but not only emotion and atmosphere, would be an interesting research topic. With the rapid progress of the affective robots, they have greater intelligence, for example, at home environment, Pepper learns together with children; based on people's actions understanding, robot plays as sociable partners in collaborative joint activities. Moreover, our developing ESRS also indicates the proposed TLFSVR-TS model could be an useful and practical way to emotional robots for emotion understanding.

13.6 Summary

On the basis of establishing a good emotional robot system, this section sets up a reasonable emotional interaction scene. Firstly, based on the above mentioned multi-modal emotion recognition method, the application experiments are carried out to verify it. Then, emotion and personal information are used to realize the understanding of emotion intention in specific situations. Finally, the application experiment of intention understanding based on multimodal emotion recognition is completed.

References

1. Y. Yamazaki, F. Dong, Y. Masuda, Y. Uehara, P. Kormushev, Intent expression using eye robot for mascot robot system. Comput. Sci. 576–580 (2009)
2. J. Hu, M.P. Wellman, Nash q-learning for general-sum stochastic games. J. Mach. Learn. Res. **4**(4), 1039–1069 (2003)
3. B.A. Erol, A. Majumdar, P. Benavidez, P. Rad, K.R. Choo, M. Jamshidi, Toward artificial emotional intelligence for cooperative social human-machine interaction. IEEE Trans. Comput. Soc. Syst. **7**(1), 234–246 (2020)
4. M. Ficocelli, J. Terao, G. Nejat, Promoting interactions between humans and robots using robotic emotional behavior. IEEE Trans. Cybern. **46**(12), 2911–2923 (2016)
5. Z.T. Liu, F.Y. Dong, K. Hirota, M. Wu, D.Y. Li, Y. Yamazaki, Emotional states based 3-D fuzzy atmosfield for casual communication between humans and robots, in *Proceedings of International Conference on Fuzzy Systems*, Taipei, pp. 777–782 (2011)
6. F. Xu, J. Zhang, J. Wang, Microexpression identification and categorization using a facial dynamics map. IEEE Trans. Affect. Comput. **8**(2), 254–267 (2017)
7. Z. Yang, S. Narayanan, Modeling dynamics of expressive body gestures in dyadic interactions. IEEE Trans. Affect. Comput. **8**(3), 369–381 (2017)
8. M. Sheikhan, M. Bejani, D. Gharavian, Modular neural-SVM scheme for speech emotion recognition using ANOVA feature selection method. Neural Comput. Appl. **23**(1), 215–227 (2013)
9. Y. Ding, X. Hu, Z.Y. Xia, Y.J. Liu, D. Zhang, Inter-brain EEG feature extraction and analysis for continuous implicit emotion tagging during video watching. IEEE Trans. Affect. Comput. https://doi.org/10.1109/TAFFC.2018.2849758
10. E.P. Doheny, C. Goulding, M.W. Flood, L. Mcmanus, M.M. Lowery, Feature-based evaluation of a wearable surface EMG sensor against laboratory standard EMG during force-varying and fatiguing contractions. IEEE Sens. J. **20**(5), 2757–2765 (2020)
11. J.C. Peng, Z. Lina, Research of wave filter method of human sphygmus signal. Laser Technol. **40**(1), 42–46 (2016)
12. A. Kolling, P. Walker, N. Chakraborty, K. Sycara, M. Lewis, Human interaction with robot swarms: a survey. IEEE Trans. Hum.-Mach. Syst. **46**(1), 9–26 (2015)
13. E. Bicho, W. Erlhagen, E. Sousa, L. Louro, The power of prediction: robots that read intentions, in *Proceedings of the IEEE/RSJ International Conference on Intelligent Robots and Systems*, Vilamoura, Portugal, pp. 5458–5459 (2012)
14. M.A. Salichs, R. Barber, A.M. Khamis, M. Malfaz, J.F. Gorostiza, Maggie: a robotic platform for human-robot social interaction, in *Proceedings of the IEEE Conference on Robotics, Automation and Mechatronics*, Bangkok, pp. 1–7 (2006)
15. K. Hirota, F. Dong, Development of mascot robot system in NEDO project, in *Proceedings of International IEEE Conference*, Varna, vol. 1, pp. 38–44 (2008)

16. K. Itoh, H. Miwa, M. Matsumoto, Various emotional expressions with emotion expression humanoid robot WE-4RII, in *Proceedings of IEEE Technical Exhibition Based Conference on Robotics and Automation*, Minato-ku, Tokyo, pp. 35–36 (2004)
17. L.F. Chen, Z.T. Liu, F.Y. Dong, Y. Yamazaki, M. Wu, Adapting multi-robot behavior to communication atmosphere in humans-robots interaction using fuzzy production rule based friend-Q learning. J. Adv. Comput. Intell. Intell. Inform. **17**(2), 291–301 (2013)
18. Blue Frog Robotics, Buddy: the first emotional companion robot (2018). http://www.Bluefrogrobotics.com/en/buddy_the-emotionnal-robot/
19. Hanson Robotics, Sophia (2016). http://www.hansonrobotics.com/robot/sophia/
20. Live Science, Meet Erica, Japan's next robot news anchor (2018). https://www.livescience.com/61575-erica-robot-replace-japanese-news-anchor.html
21. J.H. Lui, H. Samani, K.Y. Tien, An affective mood booster robot based on emotional processing unit, in *Proceedings of 2017 International Automatic Control Conference*, pp. 1–6 (2018)
22. T. Somchanok, O. Michiko, Emotion recognition using ECG signals with local pattern description methods. Int. J. Affect. Eng. **15**(2), 51–61 (2016)
23. L.F. Chen, Z.T. Liu, M. Wu, F.Y. Dong, Y. Yamazaki, K. Hirota, Multi-robot behavior adaptation to local and global communication atmosphere in humans-robots interaction. J. Multimodal User Interfaces **8**(3), 289–303 (2014)
24. M. Klug, A. Zell, Emotion-based human-robot interaction, in *Proceedings of IEEE International Conference on Computational Cybernetics*, pp. 365–368 (2013)
25. L. Boccanfuso, E. Barney, C. Foster, Emotional robot to examine differences in play patterns and affective response of children with and without ASD, in *Proceedings of ACM/IEEE International Conference on Human-Robot Interaction*, pp. 19–26 (2016)
26. Z.T. Liu, M. Wu, W.H. Cao, L.F. Chen, J.P. Xu, R. Zhang, A facial expression emotion recognition based humans-robots interaction system. IEEE/CAA J. Autom. Sin. **4**(4), 668–676 (2017)
27. L.F. Chen, M. Wu, M.T. Zhou, Z.T. Liu, J.H. She, K. Hirota, Dynamic emotion understanding in human-robot interaction based on two-layer fuzzy SVR-TS model. IEEE Trans. Syst. Man Cybern.: Syst. **50**(2), 490–501 (2020)
28. Z.T. Liu, F.F. Pan, M. Wu, W.H. Cao, L.F. Chen, A multimodal emotional communication based humans-robots interaction system, in *Proceedings of the 35th Chinese Control Conference*, Chengdu, vol. 4(4), pp. 668–676 (2016)

Index

A
Accelerated Gradient Descent (AGD), 58
Action Unit (AU), 164, 240
Action Units inspired Deep Networks (AUDN), 71
Active Appearance Models (AAM), 16
Active Shape Models (ASM), 16
ActivMedia Robotics Operating System (ARoS), 237
AdaBoost-KNN, vi, 10, 41
AdaBoost-KNN-DO, 50
Adaptive Feature Selection (AFS), 41, 42
Adjusted Weighted Kernel Fuzzy c-means (AWKFCM), 133
AFS-AdaBoost-KNN, 51
AFS-AdaBoost-KNN-DO, 43, 48, 50
AMM, 16
ANOVA, 190, 193, 206, 234
Ant Colony Optimization (ACO), 78
Artificial Neural Network (ANN), 78
Autoencoders (AEs), 58
Average Absolute Relative Deviation (AARD), 145, 207, 208

B
Back Propagation Neural Networks (BPNN), 7, 84, 123, 206, 208
Back-propagation, 26
BPA, 121

C
CAD-60, 122
Candide-3 model, 41, 43
Canonical Correlation Analysis (CCA), 92
CGD 2011, 122
CNN with Visual Attention (CNNVA), 71
Conjugate gradient, 26
Convolution Neural Network (CNN), 7, 26, 58
Correlation Coefficient (CC), 174, 242
Cross Validation test (CV test), 123
Cross-zero rate (zcr), 78

D
DAR, 70
Data clustering algorithm, 19
Decision-level fusion, 5
Deep Belief Network (DBM), 7
Deep Convolution Neural Network (DCNN), 58, 92
Deep Learning (DL), 58
Deep Neural Network (DNN), 25, 58
Deep Sparse Autoencoder Network (DSAN), 25
Dempster-Shafer (D-S), 120
Depth of Belief Networks (DBN), 26
Direct Optimization (DO), 41, 47

E
Electrocardiograph (ECG), 77
Electroencephalogram (EEG), 77, 217, 224
EMG, 224
Emotion-age-gender-nationality, 183
Emotional human-robot interaction system, 11
Emotional interaction, 215
Emotional Social Robot System (ESRS), 27, 48, 135, 163, 171, 231, 238, 242
Emotion databases, 5
Emotion expression, 190

L. Chen et al., *Emotion Recognition and Understanding for Emotional Human-Robot Interaction Systems*, Studies in Computational Intelligence 926,
https://doi.org/10.1007/978-3-030-61577-2

Emotion intention understanding, v
Emotion Processing Module (EPM), 203
Emotion understanding, 8
Empirical Risk Minimization (ERM), 136, 185
Enhanced Gabor (E-Gabor), 70
Evolutionary Algorithm (EA), 58
Exemplar-based SVM (ESVM), 68
Extreme Learning Machine (ELM), 6, 7
Eye-robot Control Module (ERCM), 203

F
Facial Action Coding System (FACS), 16
Facial expression, 217, 220
Facial Expression Recognition (FER), 57, 218
Facial Expression Recognition Module (FERM), 203
Fast Fourier Transformation (FFT), 18
Feature-level fusion, 5
Feedforward neural network, 7
Flux Approximation (FA), 42
Four-layer Restricted Boltzmann Machine (FRBM+), 70
Fuzzy Broad Learning System (FBLS), 93
Fuzzy C-means (FCM), 17, 22, 79, 163, 208, 240
Fuzzy Support Vector Regressions (FSVR), 9, 133

G
Gaussian filter, 20
Gaussian Mixture Model (GMM), 78
Genetic Algorithm (GA), 58
Gesture Recognition Module (GRM), 203
Gradient descent, 26

H
Hidden Markov Model (HMM), 78, 237
Human cognitive behavior, 218
Human-machine emotional interaction, 217
Human-oriented interaction, 9
Human-Robot Interaction (HRI), 3, 41, 162, 183
Hybrid Genetic Algorithm (HGA), 57, 58

I
IF-THEN, 198
Information - communication atmosphere, 237
Intention-behavior gap, 184
IT2FNN-SVR, 162

J
JAFFE, 29
JAFFE database, 48
JAFFE expression database, 17

K
Kernel Fuzzy C-Means (KFCM), 137
Kernel ridge regression, 26
KFCM-FSVR, 174, 175
K-Nearest Neighbour (KNN), 41, 42, 79

L
LDTP, 68
Leave-one-subject-out (LOSO), 123
Linear Discriminant Analysis (LDA), 4
Local Binary Pattern (LBP), 16
Local Binary Patterns from Three Orthogonal Planes (LBP-TOP), 92, 93
Local Zernike Moment with Motion History Image (LZMHI), 70
Logistic Regression (LR), 26, 79
Low Level Descriptors (LLDs), 78

M
Mascot Robot System (MRS), 143, 185, 203, 211, 232, 234
MEDIAN-MAX, 175
Mel Frequency Cepstral (MFCC), 18, 78
Membership Function (MF), 126, 174, 177, 242
Monogenic Directional Pattern (MDP), 70
Motion Energy Image (MEI), 17
Motion History Image (MHI), 17
Multi-dimensional emotional intention, 226
Multi-dimensional understanding, 3
Multi-modal emotion feature extraction, v, 4, 22
Multi-modal emotion recognition, v, 5, 22
Multi-modal emotional expression, 216
Multi-modal emotional information fusion, 5
Multi-modal information fusion, 224
Multimodal information fusion method, 3
Multiple Random Forest (MRF), 84
Muti-robot behavior adaptation, 212
Multi-robot cooperation, 218

N
NAO robots, 227
Natural human computer interaction, 2
Neural Network (NN), 6, 42
Novelty Search (NS), 59

O
ODFA, 42
OpenSMILE toolkit, 18

P
Particle Swarm Optimization (PSO), 42, 78
Plus-L Minus-R Selection (LRS), 41, 42, 44
Principal Component Analysis (PCA), 4, 58

Q
Quadratic Programming (QP), 185

R
Random Forest (RF), 79
Recurrent Neural Network (RNN), 7
Region of Interest (ROI), 231
Regions of interest, 26
Restricted Boltzmann machine (RBM), 26
Ridge regression, 26
RMS energy, 18, 78
Roam-robot, 183
ROBIN, 216
Robot Technology Component (RTC), 203
Robot Technology Middleware (RTM), 203
ROI region's segmentation, 17
RTM-Network, 203

S
Salient Facial Patches (SFP), 68
Scale-Invariant Feature Transform (SIFT), 5, 116
SLAM, 3, 217
Softmax regression, 25
Softmax Regression-based Deep Sparse Autoencoder Network (SRDSAN), 25, 26, 231
Sparse Coding (SC), 21, 115, 119
Sparse Coding Spatial Pyramid Matching (ScSPM), 116
Sparse Representation Classifier (SRC), 123
Speech Emotion Recognition (SER), 19, 77
Speech Recognition Module (SRM), 136, 185, 203
Speeded-up Robust Features (SURF), 5, 20, 115, 118
Standard Deviation (SD), 174, 242
Subclass Discriminant Nonnegative Matrix Factorization (SDNMF), 68
Support Vector Classification (SVC), 136, 185
Support Vector Machine (SVM), 6, 7, 42, 78, 185
Support Vector Regression (SVR), 79, 136, 162, 166, 175
SVR-TS, 161, 170

T
Takagi-Sugeno (TS), 162, 169, 240
Three-Layer Weighted Fuzzy Support Vector Regression (TLWFSVR), 8, 9, 133, 135
Two-layer Fuzzy Multiple Random Forest (TLFMRF), 77, 79
Two-layer Fuzzy Support Vector Regression (TLFSVR), 134, 162, 184, 234
Two-layer fuzzy support vector regression-Takagi-Sugeno (TLFSVR-TS), 11, 161, 163, 239
Two-Stage Fuzzy Fusion based-Convolution Neural Network (TSFFCNN), 93
Two-Stage Fuzzy Fusion Strategy (TSFFS), 92

U
Understanding Accuracy (UA), 174, 242

W
WAKFCM, 135
Weight-adapted Convolution Neural Network (WACNN), 57
WPCA, 66

Z
Zero-Crossing Rate (ZCR), 18
Zero-phase Component Analysis (ZCA), 60